Contents

CHANGING SAFETY'S PARADIGMS

Ron C. McKinnon

GOVERNMENT INSTITUTES
An Imprint of
THE SCARECROW PRESS, INC.
Lanham, Maryland • Toronto • Plymouth, UK
2007

Published in the United States of America
by Government Institutes, an imprint of The Scarecrow Press, Inc.
A wholly owned subsidiary of
The Rowman & Littlefield Publishing Group, Inc.
4501 Forbes Boulevard, Suite 200
Lanham, Maryland 20706
http://www.govinstpress.com/

Estover Road, Plymouth PL6 7PY
United Kingdom

The reader should not rely on this publication to address specific questions that apply to a particular set of facts. The author and the publisher make no representation or warranty, express or implied, as to the completeness, correctness, or utility of the information in this publication. In addition, the author and the publisher assume no liability of any kind whatsoever resulting from the use of or reliance upon the contents of this book.

British Library Cataloguing in Publication Information Available

Library of Congress Cataloging-in-Publication Data

McKinnon, Ron C.
Changing safety's paradigms / Ron C. McKinnon.
p. ; cm.
Includes bibliographical references and index.
ISBN-13: 978-0-86587-155-7 (pbk. : alk. paper)
ISBN-10: 0-86587-155-8 (pbk. : alk. paper)
1. Industrial safety. I. Title.
[DNLM: 1. Accidents, Occupational—prevention & control. 2. Safety Management.
3. Occupational Health. 4. Safety. WA 485 M478c 2007]
T55.M328 2007
363.11—dc22 2007001818

∞™ The paper used in this publication meets the minimum requirements of American National Standard for Information Sciences—Permanence of Paper for Printed Library Materials, ANSI/NISO Z39.48-1992.
Manufactured in the United States

This book is dedicated to the millions of workers who have died or have been injured as a result of accidents at the workplace because of inadequate safety measures as a result of a number of safety beliefs, myths, and paradigms.

It is also dedicated to those professionals in the safety and health business who ignored the safety paradigms and were prepared to tackle these paradigms head on and consequently made a difference to safety at their organization.

Preface

Challenging Safety's Sacred Cows

THE SAFETY PROFESSION'S EFFORTS are hampered and often rendered ineffective by certain paradigms and myths concerning accidents and their prevention. Examining these paradigms and exposing them for what they are—absolute myths—will enable industry to relook at the way it manages safety at the workplace.

Today, well into the new millennium, safety problems are still hounding modern society. Although capable of making frequent trips to space and beyond, man still can't seem to avoid accidents that kill, maim, cripple, and injure people and play havoc with our environment.

At work, each day, some sixteen Americans die as a result of work-related accidental injury and/or occupational disease. More than 4 million suffer serious injury at work each year, and a further 5 million injuries occur during off-the-job activities.

"Workplace safety still looms large in China as a serious challenge given that more than 75,000 people lost their lives in over 631,000 workplace accidents" (*People's Daily*, China).

"Over 6,000 deaths occur each year in the European Union from work-related accidents and injuries. It is largely the result of carelessness, negligence, bad management and incompetence" (European Union).

"In 2001, there were 5,300 work-related unintentional-injury deaths in the US. The cost of these deaths was $132 billion" (National Safety Council).

According to the 2004–2005 injury statistics released by the British Health and Safety Executive (HSE), 220 workers were killed during the year giving a rate of 0.7 fatalities per 100,000 workers. During the same period, 361 members of the public were fatally injured.

Three hundred and sixty thousand reportable injuries occurred, while a further 150,000 other injuries to employees were reported, giving an injury rate of 587 injuries per 100,000 employees, according to the Labor Force Survey.

The HSE also reported that 2 million people were suffering from an illness that they believed was caused or made worse by their current or past work. The report also stated that 576,000 of these illnesses were new cases in the past twelve months.

As a result of these injuries and illnesses, the HSE reports that 35 million working days were lost overall, 28 million due to work-related ill health and 7 million due to workplace injury.

On an average day, according to the fall issue of *Compensation and Working Conditions* (1997), sixteen workers are killed at work in the United States. Three are killed in collisions, and two to three are shot while doing their job. Two workers are killed by falls from elevated platforms. Each day, one worker is killed after being struck by a vehicle, one worker is killed either in an aircraft, in a water vehicle, or in a railroad accident, and a falling object kills another. On average, one worker is electrocuted every day, and another commits suicide as a result of a work-related incident. One worker is killed by being caught up in a machine or by being caught under some form of cave-in accident.

During an average year, thirty workers die from heatstroke, another thirty die from carbon monoxide poisoning, and fifteen die after being struck by lightning. Thirteen are killed by accidental shootings, and twelve are gored to death by bulls and other farm animals. More than twelve women are killed at work by their husbands or ex-husbands, according to *Compensation and Working Conditions.* Ten are killed by exploding tires, typically while inflating the tire, and during 2000 some 7 million tires were recalled by Firestone because they were rated as unsafe.

The fact that the U.S. government rated sport-utility vehicles (SUVs) hazardous because they could roll over easily at high speed was kept from the public for ten years. If one looks at the average workday and work year, it's clear that safety needs drastic improvements and that certain paradigms must be changed.

The safety profession is in a dilemma. If it's not, then it should be. All-encompassing safety and health legislation was promulgated in the United States as late as 1970 in the form of the Occupational Safety and Health Administration, and since then, everything has been done to beat the system rather than apply it in an effort to reduce accidental losses.

With the work fatality rate climbing and the number of injuries not decreasing rapidly (except where the records are fudged), it's time to admit that we have to relook at the way we practice safety in mines and industry.

We still preach and believe that *safety is first.* Meanwhile, the United States experiences 4 million disabling injuries at work each year. Why? Because we have lost sight of the truth that safety is *never* first, has *never* been first, and will *never* be first. This is but one of the many paradigms that cloud our safety vision and hamper the prevention of accidents at work.

Money, profits, production, tonnage, units—these are what is first. In modern society, the safety profession needs to face this fact. The safety profession must practice the profession so that it can put safety higher up on the agenda or as close as possible to those things that industry, manufacturing, and mining put first. How do we do this? We have to get rid of safety's paradigms that are hampering progress in safety. We must change the way we have viewed safety, return to basics, and question and challenge the safety myths. We need to get out of the safety mind-set box that ties our hands.

This book takes a realistic look at safety in the workplace, where we are and where we should be, and examines a few of the many myths surrounding the safety of workers. Management, employees, and the safety profession have to change certain beliefs and antiquated ideas to be able to put safety in its rightful position in the management of people at work. They need to change a number of paradigms and many more safety myths to be able to do that. Many positive ideas are given as to how management and the safety profession can improve its thrust for safer and healthier workplaces by identifying and dispelling safety paradigms.

Ron C. McKinnon, CSP

1

Safety Paradigms

A PARADIGM IS A THING THAT we have always believed and accepted. It is a mind-set and can easily reach the status of myth. We tend to accept the mind-set of the current paradigm, and therefore we are in danger of remaining locked into preconceived concepts. Many years ago when Alexander Graham Bell demonstrated the telephone, members of Congress agreed that they saw no future for this device but did mention that it had entertainment qualities. They were locked into thinking inside a particular paradigm. Look around your home today at the three telephones you have and the e-mail connection and the cellular phone that you, your wife, and your children have, and you will understand that the statement by those members of Congress was based on an outdated paradigm.

The profession of occupational safety is full of "safety" paradigms, and the top few will be examined. *Safety is first* is the first myth or paradigm. Safety is never first and will never, ever, be first. The best it can strive for is to become recognized as one of the facets of day-to-day life, business, and profit-making attempts. We must change the prime paradigm that safety is first, immediately, and then start changing some of the others.

Everybody Is Responsible for Safety

I have heard the paradigm that everyone is responsible for safety repeated time and time again. I have heard this from groups of safety seminar delegates. I have heard it from senior management. Employees quote this almost as rigorously as some religious chant. Once I asked the question to a group of seminar attendees, "Who's responsible for safety?" A chorus, one voice, united, replied, "Everyone!" Not so! How can *everybody* be responsible for safety? Analyze it, evaluate it, and think rationally, and you will discover that *everybody* cannot be responsible for safety. This paradigm creates a cop-out that gives the impression that line management is not responsible but that everybody else is. When it comes to safety, "not my job!" is heard often in the workplace.

Blanket Clause

This is what is referred to as a blanket clause. It is too generalized, too generic, and it is absolutely meaningless. Yet people around the world have started to accept and believe that *everybody* is responsible for safety. If everybody is responsible for safety, why then do we need management? Why do we need production leaders? Why do we need team leaders or supervisors? If one had to generalize that everybody is responsible for safety, then one could say, "Well everybody is responsible for making a profit." Some departments cannot make a profit, as their functions are staff or advisory related. How can everybody then be responsible for safety? If *everyone* is responsible, *no one* is responsible. After an accident, look for *everyone.* This paradigm statement creates a major cop-out and gives the impression that accountability is widespread and cannot be fixed.

Insult to Injury

To go further—and to add insult to injury—the latest modification is that "everybody is responsible for his or her own safety *and* the safety of their fellow worker." This unfortunately does not work because it pushes peer pressure to the limit and is, after all, an unreasonable expectation of fellow workers and team members. The following example explains the reality of asking one worker to be another's "keeper."

During a recent seminar, I asked a large, muscular, six-foot six-inch seminar attendee to stand up. His name was Joe. I then asked a short, thin, and underweight gentleman who happened to be his colleague to stand up. His name was Jack. So here, standing in the middle of the lecture room, surrounded by a hundred other delegates, we had Joe, six feet six inches tall, muscle from neck to ankle, and Jack, five feet one inch tall, slim built, and wearing glasses. Jack looked a bit timid and terrified standing in front of the audience beside big Joe.

I then said to Joe and Jack that they were responsible for each other as per the expression, "You are responsible for your fellow worker's safety." They both nodded in agreement. I then created a scenario and asked them to do the role play as would happen in real life.

I said to Jack, "Listen Jack, you have walked past, and you noticed that Joe, 'the giant,' is working on a grinder, sparks are flying everywhere, and he is working without wearing any eye protection. He is not even using the grinder shield, and sparks and wheel fragments are flying past his face. He could suffer an eye injury at any moment." Jack nodded his head. I continued, "Now Jack, because you are responsible for Joe's safety, approach him and tell him to wear eye protection."

Jack was really wrapped up in the role playing, and he approached Joe and said, "Hey Joe, shouldn't you be wearing eye protection?" Joe looked down at Jack and, ignoring that this was a role play, reverted to natural instincts and said, "Who the hell are you to call me Joe? Around here they call me Mr. Joe!" This was a natural reaction from Joe and had not been rehearsed or encouraged at all. This was a risky role play, as it was done during a public safety seminar. Jack tried again. "Joe, please, I am here for your safety. Wear your eye protection." Joe started to turn red, he sucked in air, his chest expanded, and he looked at Jack and said, "Hey, you get lost, stop trying to tell me what to do, you're not my boss!"

I stopped the role play at this stage because obviously this was true-to-life stage performing. Joe didn't like being told what to do by anybody, let alone by his colleague Jack.

Unless it is our managers or boss who tell us what to do, we tend to react negatively. How can one colleague give an instruction to or reprimand another colleague? They are on the same level in the organizational hierarchy and therefore have no authority. People don't like receiving instructions, being corrected, or being told to follow safety rules by others on the same level as themselves.

In concluding the seminar, I asked the attendees if they would have reacted the same as Joe "the giant." Ninety percent nodded their heads in agreement; "yes," they don't like being told to follow safety rules by their coworkers. They disliked being told to follow safety rules by somebody on their own level.

I concluded by asking how many of the attendees had problems convincing their wives or other family members to wear vehicle safety belts when going on a Sunday afternoon drive in the family car. No less than 70 percent agreed and admitted that they have had problems in convincing their immediate family of the safety rules. If 70 percent have problems convincing their immediate family, what percentage will have problems convincing their fellow workers (especially if their fellow workers happen to be the size and posture of Joe "the giant")?

Level of Authority

All of us are responsible for safety within our limits of authority. That means that if you have the *authority* to spend $500 of the organization's money, then you have the *responsibility* of ensuring that, within that $500 level of authority, safety exists. The higher the level of authority, the more the responsibility and the consequential accountability. If you have little or no authority, how can you be held responsible? A fellow worker has very little authority over his coworker. If he has little or no authority, how can we make him responsible for his fellow worker? It does not make sense, is unfair, has never worked, and will never work.

If you have authority to give instructions to fifty people, then you have the responsibility for giving safety instructions to those people, and you will be held accountable for those fifty should you not give correct instructions or should they not follow your instructions. You can't be held responsible for the other fifty *not* under your control, can you? This is the "I am my brother's keeper" syndrome, which may work in politics but doesn't work in safety. If a person has no authority, how can he or she be held accountable for items as important as safety? The higher up the organizational structure, the more authority people have, and therefore the more accountable they become for the safety of the people and the organization.

Span of Control

The management principle of the *span of control* states that the more people a supervisor has under his control, the less capable he will be of managing their activities. Supervisors and other levels of management are normally allocated the manpower and resources that they can manage comfortably. A span of control that is too narrow means that the resources under the supervisor's command are overmanaged. If there are too many resources and people under his control, they will be undermanaged. Span of control will also determine what authority, responsibility, and accountability a supervisor has for safety.

Top and Middle Management and Employee Responsibility

Everybody is responsible for safety within his or her area of responsibility. Top management is responsible for all the departments, people, and resources reporting to them, middle management likewise. Employees are also responsible for safety within their sphere of authority.

If an employee is responsible for an area twenty by twenty feet containing an electric arc-welding machine, he should be held accountable for that "cabbage patch," or twenty- by twenty-foot area, and the items that he has control of within the area. Asking him to carry the brunt of safety violations outside this area is truly unfair and impractical.

Twenty-Dollar Exercise

Having studied and taught management, I have found numerous misconceptions about the three magic words of *responsibility*, *authority*, and *accountability*. Using a $20 bill, I will endeavor to clarify these management concepts. They are so important in safety, yet they are so misunderstood.

Responsibility

Safety responsibility has been defined as *the safety function allocated to a post*. So, in the $20 dollar exercise, I will hand over $20 of my money to Jack for him to invest for me, thereby making him responsible for the $20. My question would be, "Jack, what are you responsible for?" and he should reply, "I am responsible for $20." Because the $20 in Jack's hand is not going to grow, I will then give Jack *authority* to invest my $20 with a bank of his choice.

Authority

Safety authority is defined as *the total influence, rights, and abilities of the position to command and demand safety*. So now that Jack has authority to invest my $20, he has total rights to demand and command some sort of return on the money. He has the influence, the right, and the ability to invest my $20 to earn me interest. For his efforts, he will receive a part, or portion, of the interests. I have now given him *authority*. However, I have given him only $20 worth of *authority*. If Jack decides to invest $200 on his own initiative, that's his problem. I have given him only *responsibility* for $20, and therefore he will be operating *beyond* his area of *responsibility* by investing the extra $180.

Accountability

After six months, I decide I need my $20, and guess what? I will now hold Jack *accountable* for my $20—not everyone because *everyone* was not given the $20 worth of *responsibility*!

I now go to Jack and say, "Jack, I hereby hold you *accountable* to return my $20." Immediately, Jack will return my $20 and say, "You have made me *responsible* for this $20. I have taken that *authority* and invested $20, and since you are making me *accountable*, here is your $20 plus $1 interest. You gave me the *responsibility* of making your money grow, and with your *authority*, I invested it with Bank XYZ and earned $1.50 interest. The 50 cents I am taking for my trouble."

In this simple scenario, Jack was given $20 worth of *responsibility*, he was given $20 worth of *authority*, and he was held *accountable* at the end of the day for $20.

By making everybody responsible for safety, you are giving a person $2 million worth of responsibility with only $20 worth of authority. By making them accountable for the safety of their fellow workers, you are giving them only $20 worth of *authority* yet holding them *responsible* for $2 million worth of *accountability*. That is unfair. This is poor management practice. This is also a weakness in the safety setup.

The conclusion and point is that you can hold people *accountable* only for things over which they had *responsibility*. They can have *responsibility* only if you give them that *authority*.

Therefore, in summary, all people cannot be held *responsible* for safety. People who have *responsibility* are the only ones who can be held *accountable* for safety. The *accountability* that you hold them to depends on their *authority*. Their *authority* determines their *responsibility* and their *accountability*.

If everybody is *responsible* for safety, it is a cop-out, and nobody is truly *responsible*.

A good safety system entails setting standards of accountability. This is where specific people, within specific levels of *authority*, are held *accountable* for their *responsibilities* within their sphere of *authority*. Ideally, in safety, single point accountability, when only one person is accountable, is necessary.

In their Health and Safety Policy, Tourvest Travel (Pty) Ltd states the following: "The Company recognizes that safety is the responsibility of everyone within the organization and is not just a function of management. Management will have specific duties and responsibilities to comply with in the spirit of company policy. Employees have a specific responsibility to take reasonable care of themselves and others that could be affected by their activities, and to cooperate with management in achieving the standards required."

Fix It! (True Story)

During an inspection of a factory, my host and I encountered the door of a switch room wedged open by a four- by four-inch timber. Between the open door and the switch gear was a metal chair, and the doors of the panels giving access to the live electrical conductors were standing open. This was an ideal setup for somebody to trip over the four-by-four and fall with arms outstretched into the live electrical switch gear.

We contacted the electrical supervisor immediately. He explained that the switch gear was overheating and the doors of the panels needed to be left open. He also explained that the door to the switch room needed to be open to allow air to circulate through the live current carrying conductors that were generating all the heat. If the system overheated, it would trip and shut down production. It took us three weeks to get this situation rectified. The organization tolerated the risk of imminent electrocution in order to ensure that the switch gear did not overheat and shut down the process.

On relating this story to a group attending a seminar, one attendee stood up and asked me why I hadn't shut the door and closed the panels, thus rendering the situation safe. I replied by asking what would have happened if I had

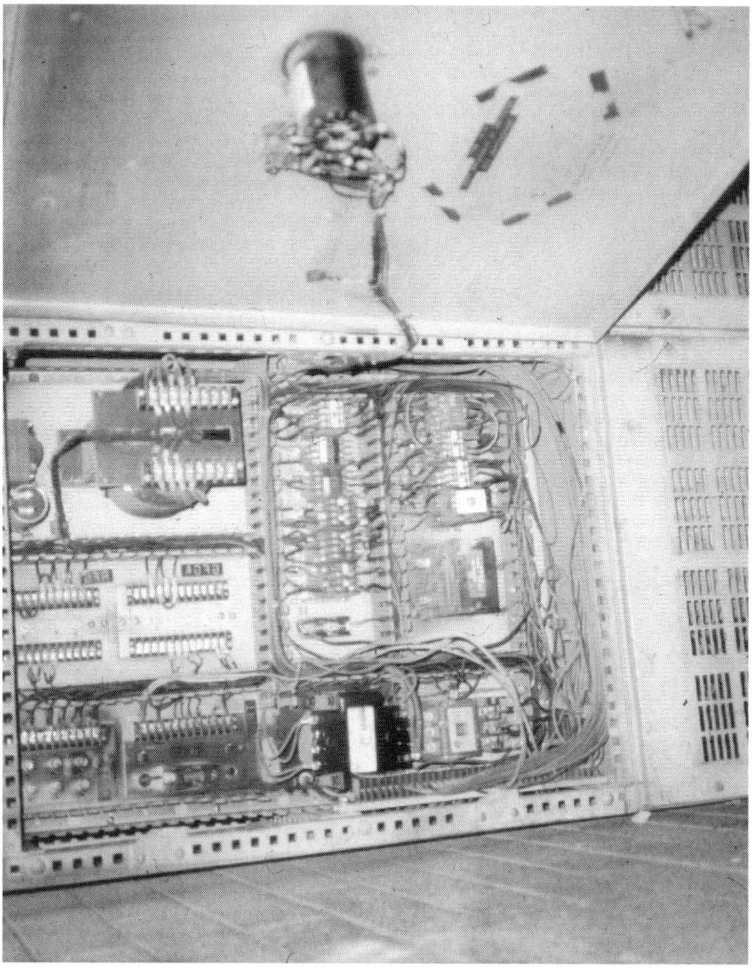

Figure 1. The open electrical distribution board.

slipped and fallen into the live switch gear and had been electrocuted? A number of people at that seminar had difficulty with my explanation.

I don't believe anybody should rectify an unsafe act or unsafe condition that is outside his or her direct responsibility and authority. I can almost see you cringe as you read this statement. Let me explain. Let's say that I had removed the four-by-four, removed the metal chair, but, while trying to close the door, I accidentally slipped and contacted the live current-carrying parts and was electrocuted. Just picture the description of the event written on the accident investigation form. I believe it would have read as follows: "*What caused the accident? Unauthorized person entered switch station without authority and performed unauthorized work on equipment, for which he had not been trained to work on or authorized to adjust, or maintain.*" This is exactly what would have happened.

So, expecting every person, at every different level, to breach across the lines of command and to transgress the managerial hierarchy to rectify hazards is asking a great deal. It is also putting that person at risk of being reprimanded should things go wrong. It would be better to take action by notifying the person responsible for the situation or area to take action.

Poor Management Systems Cause Accidents

Accidents are caused by a breakdown in the management system. Management system standards and controls fail and cause an accident. In some instances, where no system standards, controls, checks, or balances were in place, the consequence was an accident.

Management must accept their full role in the safety management arena. They have ultimate authority; therefore, they have ultimate responsibility and should be ultimately held accountable for the physical work environment and the actions of employees.

Numerous people will tell you that the behavior of a person is what causes accidents. They will justify their statement by saying that behavior is a major factor in all accidents. So what? The way people behave is the way management *wants* them to behave. Remember that management gets what it wants. If management wants safe behavior and safe work conditions, it will get them. To cop out and say, "Everybody is responsible," is abdicating and not delegating. What it really means is, "Do your own thing, but you're in trouble if you get injured."

Behavior-Based Safety

Behavior-based safety is not a comprehensive safety system but merely one element that could contribute if used correctly. As with a lot of employee behavior–focused systems, the system tends to become a "catch the person doing something wrong" scenario.

Numerous unions are formally opposing this safety strategy because they are concerned that the employers are shifting the responsibility for job injuries to the workers and are focusing on their behavior instead of focusing on the unsafe work environment. They complain that establishing procedures for observing and documenting workers' unsafe acts and at-risk behavior tends to ignore employers' mismanagement and the root cause of injury-producing accidents. They also state that disincentives to reporting injuries and hazards are brought about through rewards, intimidation, and postinjury drug testing.

Treating Symptoms

A structured system of control identifies the unsafe acts of the employees but also asks, "What were the basic causes for these at-risk behaviors?" The unsafe acts or at-risk behavior, as they are termed by the behaviorists, are only *symptoms* of a failure in the management system. With all respect to behavioral safety, it will never work unless the behaviors of management are first identified and rectified. There is always a reason why a person behaves the way he does. If a person behaves in an "at-risk manner," it is because his local supervisor is not supervising him correctly. His behavior is therefore tolerated and, as such, condoned.

As safety pioneer and expert Dan Petersen (2005) says, "Although current research shows that management is the key, many behavioral approaches focus on employees. This allows management to abdicate its responsibility for safety."

Bill Hoyle (1995) of the Oil, Chemical and Atomic Workers International, in a paper titled "Fixing the Workplace, Not the Worker," says, "The basic premise of behavior modification programs is that the primary cause of accidents is worker error. This blame-the-victim concept provides little opportunity for effective accident prevention. Behavior modification does not focus on the fundamental safety problems that we face in the continuous process industry. For example, it does not address the need to change the dangerous contractor system or the unsafe practice of running plants far beyond their safe design limits."

He further states, "It can be argued that the primary thing that has usually changed is that the reporting of injuries has been reduced. However, even if the injury rate is actually lowered, this is a very inaccurate indicator of overall workplace safety, especially in the oil and chemical industries. These plants have a very low OSHA [Occupational Safety and Health Administration] injury rate yet the number of disasters has sky-rocketed" (2).

Safety System

Don't get me wrong. Behavioral safety could play a part in the total safety system. It can have a role only once the safety system is in place and is monitored constantly by ongoing audits, checks, and balances. Sometimes safety practitioners, without an in-depth knowledge of the workings of the safety management function, tend to grab any "new approach" to safety. These are normally in the form of some gimmick or other "flavor of the month" approach to safety. Getting back to basics and recognizing near misses as more important than the actual accidents will lead to a reduction of losses. Remember that having no injury does not mean that there wasn't an accident. (No blood, no foul.)

Don't Be a Ninny

In a work situation, the macho aspect of men's personalities seems to prevail. The minute a buddy draws his attention to the safety aspects, he is immediately looked on as a ninny and a whiner. Sometimes I've even heard of people being called "safety geeks." What these unfortunate people are endeavoring to do is to take some responsibility for their colleagues' safety. In doing so, in the average workplace, they immediately set themselves up for criticism and ridicule.

The paradigm that we can work unsafely as long as there is no injury seems to prevail. Or we can work unsafely because nobody has been injured in the past. A person who does manage to convince his colleagues to follow the laid-down safety practices is indeed a diplomat, an excellent salesman, and someone who belongs in a proactive safety department.

Peer Pressure

Peer pressure is perhaps the biggest dictator of behaviors in the workplace. I do not believe that employees really make up their own minds whether to work unsafely. Peer pressure dictates the behavior. If twelve roofers are working on a roof and none of them are tied off with suitable fall protection, it is almost guaranteed that if one of them demands full fall protection, he will be frowned on by his peers and colleagues. He will probably be ridiculed during the next few jobs and receive some sort of nickname, such as "Safety Geek," "Mr. Safety," and so on.

On the contrary, peer pressure can be made to work very positively. If the team has taken to good housekeeping and accepted safety as a value, they will pressure other employees to do the same. An environment that is conducive to safe practices and conditions can bring about this positive peer pressure. This must also cascade down from the top. Remember that a manager gets what he wants. If a manager wants safe work conditions and safe work practices, he will get them. Part of the manager's achievement in safety is to use positive peer pressure rather than making people responsible for their buddy's action, which is using negative peer pressure.

Setting safety challenges based on proactive safety activities creates positive peer pressure and can be one of the most powerful methods of obtaining buy-in into the safety system. This is why motivation and applying the principles of participation and recognition are so successful in changing safety culture and getting buy-in from all levels.

Condoned Practice

Condoned practice is perhaps one of the most powerful of all peer pressures. Condoned practice means accepting what a person is doing unsafely and pretending that nothing will happen. We often feel that if our workmate is working unsafely, we should follow suit. We tend to believe that nothing will happen to us, that it's okay. The self-preservation frame of mind lures us into believing that if he is not being injured carrying out the activity, then we too will not be injured. Many an employee has been injured as a result of condoned practice.

Another scenario that lulls us into complacency is when a member of management sees an unsafe act being committed and does nothing about it. Unknowingly, this sends a very powerful message to the perpetrators that their actions are in order.

While struggling to implement a safety culture change program at a particular organization, I found all my efforts completely blocked. Neither management nor employees responded to the call to improve safety. In a large meeting, I asked the employees why they were continuing with their unsafe behavior. They said that they felt it unfair that they should work safely when management constantly committed safety violations. On questioning their reasoning, it became apparent that all the general managers, without exception, drove onto and around the factory premises without wearing their vehicle seat belts. This was a loud and clear message to employees that it was okay to breach company safety regulations.

Investigating one particular fatal accident at a mine, I spent more than seven days on the investigation and visited the accident site on no fewer than seven occasions. Only on my last visit to the site did I discover one of the reasons for the man's accidental death. The workers were erecting massive wooden poles for the purpose of running high-voltage electrical conductors to a remote part of a mine. The procedure for digging the holes had been changed before this work assignment. Instead of the holes being dug manually, a backhoe had been brought in to dig the holes mechanically so as to save on labor. The backhoe could not dig a narrow vertical hole for the pole but rather dug trenchlike openings.

The poles were then placed into these oversized trenches and, using manual labor, were partially supported by some of the earth being repacked into the holes. This provided temporary minimal support. A bulldozer was supposed to come on-site to complete the filling of the holes and consequent supporting of the poles. The temporary filling provided a very unstable support for the poles. The bulldozer was delayed, so the electricians decided to drape the electrical conductors over the insulators on top of the poles in the meantime.

During this process, one massive pole, with the electrician strapped onto it by his safety belt, collapsed. On hitting the ground, the impact caused the pole to rebound against the victim, crushing his chest cavity and causing his death.

Only on extensive investigation and site inspections did I discover what had motivated the victim to climb an unstable, insecure pole. There were numerous reasons found. In the interests of continued production, the electricians had decided to press on with the job despite the poles being insecure. One of the reasons for this was that they had waited for more than eight hours for the bulldozer to fill and support the poles.

An interesting fact was that the standard, condoned practice was that once the poles had been placed into the traditional, narrow, hand-dug holes, the workers climbed them and continued with the pulling of cables onto the conductors. The narrow holes would have supported the poles even without being filled, and there was a low probability of their falling under the circumstances. The same condoned practice of wiring the noncompacted poles had continued in this instance. Many of the poles had been successfully wired. The electricians had gotten away with climbing and wiring seven of the twenty poles. Unfortunately, the pole the victim was on was less secure than the others and pulled out of the ground while he was on the top feeding the conductors through the insulators.

The obvious condoned practice was the climbing of unsupported poles, and this created a precedent for the electricians to carry on with the job. Unfortunately for the victim, the pole he was working on collapsed during this procedure and killed him.

The change to the method of digging the holes and the condoned practice of climbing unsupported poles contributed to this man's death. I asked the question, "Why did the other poles not collapse when the other electricians climbed them?" The only answer I could find was that luck had played a role in this accident and its devastating consequences. Sometimes one can get away unscathed if lucky enough. The other electricians were just lucky.

Summary

There are numerous paradigms, myths, misnomers, misleads, and general cover-ups in safety. I believe that occupational safety and health improvement and promotion schemes will never, ever, succeed as long as these mind-sets remain.

These myths have been so embedded into our subconscious minds that we believe them. They are posted year after year on safety notice boards and are talked about at safety conferences, and the attendees return refreshed and brainwashed believing these stories. It's time to change these paradigms. It is time to admit the truth. It is time to expose these safety myths for what they really are. To make *safety first* more reasonable, we must examine, revisit, and get rid of these safety paradigms.

2

Paradigm 1: Safety Is First

THERE ARE A NUMBER OF SAFETY PARADIGMS to be discussed, the first one being that "safety is first," which is the biggest paradigm of them all.

What Is First?

In reality, what is first in a country's economy, in the mining industry, the manufacturing industry, the service industry, and any other type of business? Money is first. Profits are first. Production is first. Employee attendance, dedication, and output are first. Money and profits will always be first. Return on shareholders' investments is first. Making a bigger profit is first. Tonnage, number of units, quality standards, being "better than the competitor's product"—these are all what is first. Bigger, better, faster selling—that's what's first.

Where does that leave safety? Way back at the end of the line, last in the line. Pause for a moment here and ask yourself about your own experience as either a manager, an employee, or just the average man in the street. Where does safety feature? Is it first, is it close to first, or is it last? Everybody says "safety first." Safety is seldom first. Safety is often last—if it features at all.

As Richard Fulwiler (1997) says, "Safety First! Is this an empty, time-worn slogan or a forceful admonition to senior line management that they heed relentlessly?" He continues by saying, "It is unrealistic to think that senior line management will day in, day out, place safety above all else" (1).

In his Internet newsletter, Fulwiler makes the point that safety must be an integral part of line management's responsibility. He believes that safety should rank in the same order of importance as production, quality, and costs. "Line management needs to be taught, or convinced, that safety is not first but rather an equal—held to the same levels as production, quality and costs." His first step to the four steps of ownership is "Get off the 'safety first' kick if that is your current mindset!" (1).

Management

A special word of dedication is opportune at this stage. I personally pay tribute to the men and women who have a title that falls under the general heading of "management." This may be team leader, supervisor, group leader, foreman, *induna*, *bwana*, *hefe*, chief, subchief, monitor, or any other title that describes someone who has to manage the activities of other people. These leaders deserve sympathy. The bad news is that they are in the toughest business in the world. To aspire to a level of management is every potential leader's ambition.

I personally aspired to become a manager. I became a manager. I became the youngest regional manager (at the time) in the history of the organization that I worked for. I did what I thought was managing for a number of years until I woke up to reality. I realized that managing was exceedingly difficult. It involved dealing with personalities, domestic problems, emotional problems, political problems, cultural problems, language problems, and all the rest. If you are a manager, I know that you have these problems, and therefore I pay tribute to you and sympathize with you at the same time.

Unknown Territory

As far as safety and the prevention of accidents are concerned, I know that most managers have no formal training whatsoever in this science. Safety is therefore really unknown territory to them, and what we are all scared of is unfamiliar territory. This is why management has not embraced safety as part of their management function.

Because of this fear of the unknown, management uses a number of methods to barricade themselves from this unknown entity called safety. That is why they often use meaningless cop-outs such as "Everybody is responsible for safety," or they say, "Well the safety department must look after the safety, I'll run the business." Because of this, Mr. and Ms. Manager, safety is seldom placed first.

Training

I have interviewed almost a thousand managers in eight different countries. The questions asked concerned the formal safety education they had received during their associate's, bachelor's, or master's degree courses. Only one manager admitted that he had received safety training. When I questioned him, he said that he had attended a two-hour class on safety during his master's degree. This creates a dilemma. Safety is a management function. Safety should be part of a manager's function of planning, organizing, leading, and controlling. If safety is not taught at universities and management school, how can managers practice it? All echelons of management should accept the following facts:

1. Safety is a line management responsibility and a management function.
2. If line management has not been schooled in formal techniques of safety *guiding*, *educating*, *training*, and *motivating*, I suggest that all levels of management get some formal safety training now.

It is not sufficient to assume that "safety is first" because safety is *never* first and will never be first as long as the captain of the ship doesn't make it a priority. Safety knowledge is the first step in this process.

Do Managers Know about Safety?

Do managers really understand the safety management science? Perhaps one of the main reasons why safety is never first on the agenda but is always said to be first is that managers do not know or understand the safety management process.

College Education

It is disturbing to note that in formal management training at colleges and universities, whether it is a bachelor's or master's degree, the subject "Safety Management" does not appear on the course curricula or syllabi.

The existing system of management development, education, and promotion creates a failing in the development of the average manager. Most foremen, supervisors, and team leaders have progressed through the ranks and have not had the opportunity of formal management training. Most are promoted because of long service or because they are good at what they do. If they have not been trained in management skills, what about their safety skills?

At one large smelter, a major program of management training had been started. At the same time, I was recruited to bring about safety culture change and implement a world's best practice safety system. During my first meeting with the chief executive officer, he excitingly told me about the 400 line managers who were to embark on this in-house four-week management training program. My first question to him was, "And how much safety does the course contain?"

He was visually shocked by my question and more so by the realization that the entire management team were being trained in management skills but that safety management skills had been omitted from the course syllabus. I proposed adding an eight-hour session on modern safety management onto the end of the four weeks to which he agreed. The training proved to be a great success, as none of the 400 candidates had received any formal safety and health training at all during their careers. I also presented the same courses to members of the management executive and, with this thrust from the top, middle, and lower management levels, managed to achieve a major safety culture shift in one year.

Functions of Management

The four basic functions of management, according to Louis A. Allen, are, *planning, organizing, leading,* and *controlling*. Safety is a controlling function. Safety entails managing the identified risks of the business, identifying what work needs to be done to reduce the risks, and then setting standards of compliance. Once the standards of measurement have been determined, management then set standards of accountability for safety actions.

Ongoing measurements in the form of safety inspections, reviews, and walkabouts are then integrated into the day-to-day management. Periodic evaluations, or audits, against the laid-down standards should be part of the management agenda. This is now putting safety management first. Once there are deviations from the laid-down standards, corrective actions are initiated. Part of good management is also commending and praising where conformance to standards is encountered. This is the safety management activity in a nutshell.

Lack of Knowledge

Managers have to manage people, assets, materials, accounts, and the organization's reputation. They have to deliver a bottom line to satisfy the board of directors and shareholders. Safety is also lumped on their shoulders as a major responsibility. Is this perhaps fear of the unknown that causes management to ignore proactive safety activities and try and delegate these down the line and specifically to the safety department? The safety department, looking for recognition, wrongly accepts this responsibility, and therefore safety becomes a safety department activity and responsibility. The safety department is *not* the management of the organization and can never, will never, and should never manage the safety for management.

A Friendly Reminder

This is a friendly reminder to all the safety, health, and risk management departments. Remember this: management is responsible and accountable for the safety of the organization. Only management has that authority. A safety and health department does not carry the responsibility for safety. No safety department, in any country, has the authority that is necessary for the responsibility and accountability for safety.

Safety Department

If your department has been appointed as "responsible for safety," approach management immediately and get the department's objective rewritten. The department is responsible for the *guiding*, *educating*, *training*, and *motivating* of all levels in the techniques of health and safety. Safety coordination, direction, and advice are the safety department's functions.

Opportunity

This is a major opportunity for the safety profession in the future. More than likely, the management team does not know about professional safety. What they know is what the safety staff has taught them. What they do not know is what safety professionals still can teach them. The safety department will manage to improve safety and change the culture only if they educate managers as to what safety management is all about. Constantly reporting to management on how many or how few people are being injured at work is not safety and will not educate them in safety management techniques.

I once encountered a safety manager in his office almost in tears. I asked him what was wrong, and he said that *his* injury frequency rate was rising steadily month by month. He looked extremely dejected, worried, and distressed. I sat him down and told him a few home truths.

The first thing I told him was that the injury rate was not *his.* I told him that the injury rate belonged to management. I explained to him that only managers get what they want. The safety department does not get what they want. I also explained to him that the rising injury rate was a clear indication that management was abdicating their safety activities. Rising injury rates are conducive of poor management control.

This particular safety manager had owned the responsibility and accountability for people being injured. So many safety departments try to control the degree of injury, to make their department and management look good. That's why safety does not succeed: the responsibility is often abdicated.

Bending the Rules

If the number of reportable injuries is the only measurement of safety that an organization uses, then cooking the books to make everyone look good seems to be condoned practice. Let's have a reality check here. Often the safety department is made responsible to reduce the number of injuries. This they cannot do, so they often resort to bending the rules to achieve their impossible task.

As Bill Hoyle (1995) so rightly says, "Workplace safety is measured by only one statistic, the OSHA recordable rate. Based on this statistical yardstick, continuous process industries continue to be among the safest industries in the country and are getting safer. Many plants have celebrated working millions of hours without a lost work day accident. While all eyes are on the OSHA recordable rate, releases of hazardous materials, fires, mechanical breakdowns and near misses are not included in the safety statistics" (7).

Opportunity to Minimize Loss

I have managed to convince a lot of managers that maybe their efforts are wasted by their trying to produce *more* articles, process *more* tonnage, and make *more* profits. What happens in here is that one tends to lose sight of the money slipping out the back door in the form of the cost of accidental injuries, property damage, and business interruptions. Fires and explosions are the most costly of all accidents and could easily wipe out a company's profits for a number of years.

Safety is an opportunity not to maximize profits but to minimize losses. Unfortunately, safety accounting is not as easy as other forms of accounting.

The direct costs of accidents, the indirect costs, and the totally hidden costs are spread out, diluted, and hidden. It is often difficult to calculate accidental loss costs and then present them to management as an accurate figure.

The immediate cost of a severe injury is not felt by an organization, as there is a bridging period during which workers' compensation and insurance pick up the expenses. Increased workers' compensation may occur only a few years down the line. By then, this increased assessment cannot be directly attributable to the accident. The many totally hidden costs, such as the harm to the organization's reputation, the slowdown of people as a result of their having seen a colleague being hurt, the inconvenience of having to order or repair machinery and equipment, are all difficult to tally. Often they are hidden in the normal day-to-day cost of maintenance.

Most safety people fail because they use the humanistic approach. Regrettably, managers are in business for the money and not for the humanitarian reasons. The shareholders are in the business for money. Each employee gets out of bed, drives to work, and does an eight-hour shift because of the money. No other reason. Why should safety be different? Trying to sell a structured safety system on the pretext that it can reduce pain and suffering won't work because it is not management language. Talk money, and management will listen. Accurate accident costing may be the greatest safety education your manager gets.

Costing Exercise

The next time your organization experiences a property damage accident or an accident that causes business interruption, do a costing, even if this has to be estimated. Remember that most managers review safety only as something that injures people. Few of them equate property damage or business interruptions as accidental losses. Present this to management and explain to them exactly how much this undesired event has cost the organization. At the same time, present a cost-benefit analysis and show how little it would have cost to prevent the event occurring in the first place. If your accident investigation is a good fact-finding exercise, you could also point out to management where weaknesses in the management system affect the profits of the business.

I have personally presented numerous seminars and safety training courses where people have been bored to tears when discussing injury, pain, suffering, and possible death. The minute I start equating unsafe acts with a loss of personal financial income, the attendees immediately prick up their ears and pay full attention. When talking about the financial implications of a failure of the management system, management also pays attention. Safety is a management function because it involves losing or saving money. That's why safety should be the first priority or at least among the top ones.

President's Message

Gerald F. Scannell (1999), president of the National Safety Council, lists ten obstacles to workplace safety. The number one obstacle, according to Scannell, is America's obsession with the bottom line. If American workplaces are obsessed with the bottom line, what a wonderful opportunity for safety to rally and help preserve and improve the bottom line by reducing and eliminating accidental losses.

Not only is there a saving to be made by accident and loss reduction, but there is also a bonus to be had by conforming to legislation. Fines for noncompliance to MSHA (Mine Safety and Health Administration) and OSHA (Occupational Safety and Health Administration) regulations (and others) are overwhelming. In some instances, fines exceed $1 million. It is common to read of fines in excess of $100,000 due to as few as fifteen citations. I believe this threat can be turned into an opportunity by making safety the first priority.

The 1989 Phillips Petroleum explosion in Pasadena, California, decreased the company's net income by $431 million that year. The resultant twenty-three fatalities and 314 injuries were only the direct costs of the accident.

The 1990 Channelview, Texas, chemical plant explosion cost the company seventeen fatalities and $3.5 million in fines. In San Juanico, Mexico, a series of liquefied gas storage explosions killed more than 500 people, injured 4,000, and resulted in $5 million in claims.

A final costing of the Chernobyl nuclear plant meltdown disaster will never be known, but the more than 250 directly related deaths, the evacuation of citizens from surrounding areas, and the encapsulation of the damaged reactor in a concrete coffin cost more than $228 billion.

The fourth obstacle given by Scannell is that managers are shortsighted and look only at the direct cost of safety without considering the long-term payback. Apparently, they view money spent on safety improvements as expenses rather than investments. What is the reason for this? Perhaps we have not done a good enough job of selling (28–31).

Investing in Safety

In *BHP Review*, an article appeared reporting a safety return of $9.40 for a $1 investment. "It's an enviable figure," quoted the article. "Any accountant or financial manager would be rolling their eyes. You would be fortunate for thinking that the only guarantee of such a high return would be on 'Lotto.' However, the 800% return is one that BHP Engineering claims it is reaping through avoidable costs by investing in safety" (BHP Engineering 1998, 14).

Managers, Have You Lost Contact?

According to A. A. Buys (1994), head of management services of the Alberton City Council, South Africa, management isolation occurs when a chief executive or manager is isolated from his worker corps as leader, with the result that he loses contact with the activities of the organization and the situations in his command order. This can be catastrophic for anyone in charge of a project or business, as that person is also in charge of safety.

One of the unique characteristics of management isolation is its subtle, imperceptible appearance in the organization. More and more accidental losses could start occurring. Housekeeping and order could worsen. When a manager realizes that he has lost contact, the process is often irreversible. In fact, he can be the cause of it through his own natural behavior. According to Buys, the same behavior that created the situation prevented the manager from seeing the actual facts through the defenses he had built.

A lack of communication is the most common cause. Many management consultants say that we have been working at communicating from the top down for too long and relying on the information to be passed from the supervisor down to the foreman and then to the employee. They believe communication based on this style is practically impossible.

Safety communication is traditionally downward. Another factor that leads to management isolation is the negligence of managers to go to the workplace and communicate sincerely with workers about problems. Safety and housekeeping inspections aren't viewed as part of his duties, and so he loses touch with what is going on at the point of action, according to Buys.

The manager is sometimes inclined to concern himself with "important matters" and to rely on others to be his eyes and ears. Safety is delegated to the safety department, which gives management only the good news and quotes improving figures. It is often due to sheer laziness and also because communication is not an easy process. He gets bogged down in his own daily routine and is unaware of the fact that the informal organization is formulating its own objectives that are, in many cases, not in line with those of the organization itself.

Buys believes that incomplete delegation can also lead to a break with reality. By merely delegating without follow-up, information on the cause of events becomes incomplete. Often work is delegated without the delegates being trained or given authority to carry out or maintain that which has been delegated.

According to Buys, some managers maintain an "open-door policy" but fail to give subordinates emotional security to communicate openly. The status-conscious head that isolates himself from his workers will rapidly become exposed to management isolation. His personal glorification comes first.

The uncertain manager (one who believes he can lose face), if he wants to know more about the implementation of a certain project, is busy losing contact with reality. Buys further states that it is only logical that he does not want to become intimately entangled with detail, but he has to maintain a healthy balance so that subordinates do not feel that their judgment is doubted too frequently. A channel for spontaneous interaction must be created.

Usually, an executive manager has an opportunity to implement communication channels when he is appointed. He can take the necessary steps early in his career to inform his workers that he has no intention of being cut off from them. Regular informal chats with all workers individually or in groups can be very productive, according to Buys. Although no problems are solved or definite answers supplied, subordinates are assured of an interest in their views and that this will be taken into consideration during decision making.

Steps to Improve Contact

Buys states that prevention or reduction in management isolation requires dedication and adaptation in management style. Some of the following basic steps may be followed:

- Cultivate a genuine desire to listen to bad as well as good news
- Leave your office and visit workers and their activities
- Develop the art of listening in an unprejudiced way to the right people; that is, choose the most enlightened person on the particular subject as your source
- Ask questions about safety checks, inspections, and controls
- Inspect for housekeeping hazards
- Question supervision as to their involvement in the safety activities
- Praise supervision and employees for good safety standards

"A manager's effectiveness is hampered if he is out of touch with reality, and his decisions concerning planning, change, and problems can be detrimental to the organization. He is most effective when he is in touch with the 'coal face' of his business," says Buys (1994, 4).

The chief executive and his managers must establish an organizational climate not only within which the workers can purposefully and effectively pursue the objectives of the organization but also within which they can satisfy their personal pursuits and needs to a large extent. Buys provides some solution by saying that this can be achieved by delegating authority to the lowest possible management levels, developing leadership, and extending teamwork.

"Complicated management systems and procedures are often developed and implemented at great expense, because management thought it was the right thing to do," says Buys (1994, 5). However, how does the worker corps accept this scientifically designed system, one that was developed by experts far removed from the practical work situation? He asks, "If workers are involved in the specific system, participate and contribute inputs during its development, any resistance to change in working conditions brought about by the new system, is minimized" (5). Buys asks the thought-provoking question, "Does your management perhaps suffer from management isolation?" (7).

Management Questionnaire

To test your managerial commitment and involvement in the safety program, answer the following questions truthfully and honestly:

1. Do my employees report hazards to me on a regular basis?
2. How can I ensure that the employees know that I care about their safety?
3. Do my employees show signs that they doubt my total commitment to safety?
4. Do my actions show that I am committed to safety at this plant?
5. Am I personally visible at safety meetings, accident investigations, and other safety activities?
6. When I notice a hazard, do I do something about it or report it?
7. Have I recently had discussions with workers on their attitude and actions toward safety?
8. Who is most likely to be the next of my group to be involved in an accident?
9. What have I done for safety today?
10. Where is the next accident most likely to occur? What can I do about it?
11. Have I ensured that all the necessary resources are in place to ensure a safe, neat, and healthy work environment?
12. Is safety *first* at this plant?
13. Am I convinced that *all* reportable and recordable injuries are in fact being reported?

Safety Performance Measured

While on the subject of management, let's ask the question, "How is safety performance evaluated?" I can almost guarantee that the answer to this question is, "By the number of injuries." Good! Another answer will possibly be the company's disabling injury frequency rate or the number of reportable injuries or even the disabling injury severity rate, workers' compensation rate, and so on.

These are the poorest measures of safety performance. If this is how safety performance is gauged, then we are barking up the wrong tree. These are measurements of failure or measurements of consequence. Further reading will explain exactly what safety performance is and how it should be measured. It will also explain how safety performance based on injuries is also one of the biggest cover-ups and cop-outs for safety.

Safety Organizations

Most safety organizations have their failures and their successes. As we progress into exploring why safety is seldom first, it will appear obvious that a number of safety organizations have failed to put safety in the front row. These safety organizations have been at the helm of safety for some fifty or sixty years and have also fallen victim to postcontact (after the accident) activity promotion rather than precontact (before the accident) promotion. They have also failed generally to convince management of its role in safety. Numerous safety organizations have failed to make the profession of safety a profession per se. They have lost the lead and have allowed the psychologists to take over the role of safety leaders.

Is Safety Really First in Safety Organizations?

I have known of safety organizations that have allowed smoking in their building. I have worked with safety organizations that allow their employees to drive company vehicles without wearing their seat belts. There are still safety organizations that don't practice what they preach yet stand at the podium and shout from the treetops, "Safety is first." I have questioned numerous safety organizations and asked them, "What is your safety system? What system, standards, and controls have you got in place, and how well are you faring?" They have been floored by my questions and have never invited me back to their organization. I wonder why.

I recently attended a large conference attended by more than 3,500 safety and health professionals. To access the hotel from the conference center, one had to cross a main road that ran between the hotel and conference center. I was standing with thirty other safety delegates (fresh out of a lecture on safety) at the crossing, which was regulated by a large pedestrian sign that flashed in large, clear, bright letters whether it was safe to *walk* or *don't walk*. All thirty safety congress attendees ignored the basic rules of safety and crossed over the road despite the large flashing light warning them *don't walk*.

Attending a state governor's safety and health convention as a speaker, I was appalled to see that of the twelve passengers in the hotel-to-conference shuttle bus, only three of us had taken the time to fasten our seat belts. Safety first is not applied even at safety conferences by delegates who try to preach safety to others and wonder why they are unsuccessful. To promote safety, one has to set an example in safety, and that is a fact, not a myth.

Safety Professionals

To my dear friends and colleagues in the safety profession, this book is an effort to put our profession in the right perspective. We need to expose the myths that make our task almost impossible so that a change can be made as to how accidents can be prevented. The shortcomings and paradigms that exist in the safety profession must be viewed with an open mind. After studying these, you may want to revisit your profession and the way you go about business.

I will be discussing safety professionals later on, but right now, at an early stage, the best advice I can give to the safety professional is that if you want to be a true safety professional, help change these paradigms. Take a stand and be realistic about the paradigm "safety is first" and the other myths surrounding our profession.

Join me in showing the rest of the profession how putrid the expression has become and how real it is that "safety is never first."

It is appropriate here to reproduce the *Principles and Canons and Proposed Professional Code of Conduct* of the American Society of Safety Engineers (1999):

> Fundamental Principles:
>
> Number 2. "Serve the public, employers, employees, and clients with fidelity, *honesty*, and *impartiality*."
>
> Number 4. "Avoid conflicts of interest and *compromise of professional conduct*."

We will discuss numerous so-called safety professionals and recall a number of activities that safety professionals do and also numerous conversations, discussions, and interviews I have had with them over the past thirty-three years.

Board of Directors

This is a very simple test for the chairman of the board. Where is safety on the agenda for the forthcoming meeting? Is occupational safety and health a fixed item for discussion at board meetings? Does the board of directors have safety on the agenda of their bimonthly, quarterly, or annual meeting? If safety is not on the agenda, it will not get the attention and action it deserves from the board of directors. Safety policy originates from the boardroom.

If safety were first, then surely we wouldn't be injuring and killing the hundreds of thousands of people that we do every year. People are injured and killed on the roads, at work, at home, and even in public places. Lawyers base their business on accidental injuries and deaths and make a fortune by seeking a guilty party and making that person or organization pay. More time and money are spent on safety litigation than on eliminating hazards and thus preventing accidents.

Statistics

If safety really is first, then why do such horrendous injury and fatality statistics appear each year? According to the National Safety Council (1993), there were 83,000 accidental deaths during 1992. During the same period, some 17 million disabling injuries were reported and recorded. Two million were as a result of motor vehicle accidents, 3.3 million as a result of work injuries, 6 million as a result of home accidents, and 5.8 million as a result of public accidents.

These statistics are only for the United States. If we combined statistics from other countries, the figures would be mind boggling. Can we still say that "safety is first"? If we consider the number of people killed and injured as a result of unintentional (accidental) events, we do have a safety problem. Going around shouting that safety is first is really hypocritical. This is why "safety is first" is a paradigm.

Cost of a Life

Managers, employees, safety representatives, and safety coordinators often say, "You *can't* put a cost on a life." Well, you can! The British do it, the Americans do it, and everybody else does it. What receives priority? Money does, not a person's life. The National Safety Council (1993) gives the cost of a life, in the form of an accidental death, as $780,000. In the council's version, based on the 1998 statistics, the cost of a death is $980,000. In terms of high-pressure production, competition, international trade, and business, this is a trivial amount.

Do you really believe that a loss of $1 million, which may or may not occur, will deter management from achieving the end objective of their business? Certainly not. It is business as usual. Even though there were 8,500 people killed at work in 1992, business and industry are still not motivated to put safety anywhere near their top ten priorities. Every sixty-two minutes during 1998, an American worker was killed, and six were disabled so badly as a result of accidents that they could not return to work for a shift or longer. So much for safety being first.

Grant Purdey (1996) gives another example of how costs of fatalities and injuries precede the cost of safety initiatives. He says that a suitable evaluation of statistical fatalities for decision-making purposes is attentively suggested to be in the region of 2 million pound sterling (USD 4 million). More safety-conscious companies would be justified in using evaluations up to 50 million pound sterling (USD 100 million). The cost of accidents which could have been averted by safety measures includes the following:

- The value of life of people killed: As described previously, a value of 2 million pound sterling (USD 4 million) is tentatively proposed here.
- The cost of hospital treatment, lost production, and injury payments to people injured: Based on various studies, the costs of severe, moderate, and mild injuries are taken to be 7, 0.7, and 0.14 percent, respectively, of the cost of a life.

Purdey (1996, 418) quotes the authority and the value of life as follows:

- Department of Transport, United Kingdom—0.66 million pound sterling (USD 1.32 million) (1992 prices)
- Health and Safety Commission, United Kingdom—2 million pound sterling (USD 4 million) (1991 prices)
- British Petroleum (petroleum and chemical industry)—0.2 to 3 million pound sterling (USD .4 million to 6 million) (1989 prices)

So what am I saying here? What I am saying is that safety cannot be, is not, and will never be first. Even in preventing a fatal accident, the cost of preventing the fatality is what justifies whether money is going to be spent to prevent the accident.

Purdey (1996) gives an example by talking about cost-benefit analysis and says,

> Cost benefit analysis involves a comparison of the cost of risk reduction measures with the risk-factored cost of the accidents prevented. . . . To make this comparison, the costs and benefit must be expressed in common units. Traditionally this has been in monetary units, which involves the contentious valuation of risks to *human life* in monetary terms. (419)

He quotes David Raynor, director of safety, Railtrack, on their decision not to install automatic train protection (ATP) on the British Rail Network, as required after the Clapham Junction inquiry. Rayner said that ATP would be horrendously expensive to install at a cost of 14 million pound sterling (USD 28 million) for every life saved. "No final decision has been made, but we are beginning to conclude that there is not even a case for partial installation."

So clearly, the decision in this case was that at 14 million pounds (USD 28 million) to save a life, it was not a bargain. Lives in this particular instance were only worth about half million sterling (USD 1 million), so why invest 14 million pounds (USD 28 million) to save a life? Therefore my argument: safety is never first, can never be first, and will never be first. Money is first. Let's remove this cliché from our vocabulary now because safety is never first—it's a paradigm.

Mining Industry

The following article by Fiona Leney (1994, 14) clearly emphasizes that production comes first. "Three miners dead, but production will not be affected!" is the title. According to Leney, production would not be materially affected at Randgold's *Blyvooruitzicht* mine despite the rock fall, which killed at least three miners, the mine's management said.

> Production "Unaffected" at *Blyvoor*
>
> "It is quite a scattered, old mine, with faces (working areas) several miles away from the accident site, so mining is continuing elsewhere," said a source at the mine that wished to remain anonymous. (I wonder why?—Editor)
>
> She reported that the mine's main shaft, which produces about 700 kg (1,600 lb) of gold a month, was closed after the accident. While drilling had resumed, blasting would be suspended until the two miners who are still missing were found, said the spokesperson.
>
> *Blyvoor*'s share price reflected dismay at the accident and worries over disrupted production. It slipped about 7 percent, or 50c, to R7 (USD 1) the next day.

"There will be a slight hiccup in production for a day, but I have no doubt that full production will resume quickly," said the anonymous spokesperson.

The dead miners were covered by group life insurance and the mine would not be liable for large compensation payments, the source said.

A mining analyst, who did not wish to be named, said he doubted *Blyvoor's* share price would suffer seriously from the accident. In production terms, he said, "The rock fall was negligible."

Department of Energy Admits Weapons Work Led to Illness

Even the U.S. Department of Energy (1999) admits that contract employees who became ill while constructing nuclear weapons would be compensated. "It's time to stop spending money litigating against these workers and focus our efforts on getting them the help they need," said Energy Secretary Bill Richardson at a news conference (25).

Exposure to beryllium dust, which is a metal that is a component in nuclear bombs and missiles, can lead to permanent lung damage. Serious effects usually do not surface for ten years or more. The Department of Energy estimates that out of more than 26,000 workers exposed to the toxin, about 120 have been diagnosed with chronic beryllium disease.

Military

Russell Carollo (2000) states that the military often let unsafe aircraft fly on, often with dire results. He refers to faults in helicopters and planes that are ignored, sometimes for months. According to Carollo, the true aircraft accident record is never revealed to the public, and the accident rate of unaccounted accidents increased from 1.6 to 2.3 percent in 1997.

Often, in-flight emergencies are traced to incorrectly installed parts, engines that have been overhauled incorrectly, and other mistakes made by mechanics or their supervision.

Alaska Airline Knew of Problems before Fatal Crash

The January 2000 Alaska Airline crash took more than eighty lives, yet two years before the crash, the Federal Aviation Administration had already cited the airline for incompletely documenting maintenance work, according to the *Arizona Daily Star* (2000).

In March 2000, sixty-four Seattle maintenance mechanics complained that they had been pressured by their managers to hurry repairs and cut corners. A San Francisco grand jury is also looking into allegations that Alaska Air falsified maintenance records to rush planes back into service, according to the *Arizona Daily Star* (2000).

An ex-Alaska Airline mechanic, appearing on national television, told the public that his supervisor often ignored his recommendations concerning the safety of components and systems. He recalled how maintenance documents that he had completed had been altered to allow unsafe equipment to be placed back into service. This mechanic was the only one who had the courage of his convictions to speak up and reveal the safety malpractice, and he did this before the fatal crash of Flight 261. Needless to say, he was placed on "administrative leave" until further investigation, and he may well be fired. So much for the empty statement "Safety is everybody's responsibility." I commend the maintenance mechanic's courage for speaking up. If more people would speak up before the event, we could place safety first and prevent thousands of needless accidents.

Product Recall

If safety is first, then surely there will be no product recalls. Safety in this context applies to undesired events occurring during the conception, design, manufacturing, distribution, and servicing of consumer products. Product recalls have become the joke of the safety world. Product recalls are the most expensive reasons why safety should have been considered. If safety (the control of accidental losses) had been put first, the recall would not have been necessary.

The main reason for the numerous products recalled is that the products pose a threat to the *safety* of the end user. In a number of instances, the end users are children. The Consumer Product Safety Commission (1998) has listed a number of product recalls. By reviewing this list, one gets an idea of how safety during the design and manufacturing stages was not put first and how the faulty device clearly puts the user at risk:

- 2,500 miniature rocking chairs for children were recalled as a result of the lead content in the paint used to paint these devices.
- 108,000 high-chair gym toys were recalled because a defect was found.
- 100,000 toasters/convection ovens were recalled because of faulty electrical cords.

- 800,000 infant car seats/carriers were recalled because they could collapse, causing the child to fall.
- 49,000 crib mobiles were recalled because of a hazard.
- 1.2 million detail sanders were recalled because they were faulty and could prove dangerous.

When one looks at costs of nonsafety, it is clear evidence of what poor quality, caused by undesired events (accidents), can cost. Some other recalls quoted were 5,000 climbing harnesses and 50,000 outlet converters, both of which had inherent faults and could be hazardous to the end user.

Millions of passenger vehicle tires have been recalled, having caused numerous deaths as a result of tread separation. Unfortunately, more than thirty deaths had to occur before some action was taken.

Recall Costs

Take a minute and do a quick costing of how much it would cost an organization to recall more than 1 million of its products that have been sold in the United States, and possibly exported to other countries as well. Once you have calculated this figure, consider what the costs would be of possible litigation brought against the manufacturers who have admitted having manufactured and sold faulty products.

Quality versus Safety

Faults in products are caused by poor quality. Poor quality has the same roots as poor safety. A breakdown in the management system has occurred. Fortunately, quality receives priority far more frequently than does safety. Yet the roots of safety are the same as quality and production but are, unfortunately, always viewed as separate components. Integrating safety, production, and quality is the finest decision any manager can make. Because of the importance of this statement, I will repeat it: integrating safety, quality, and production is the finest decision any manager can make and will make him a successful manager.

Conclusion

Louis A. Allen says, "Minimizing loss is as much improvement as maximization of profits." That is where safety could feature as first priority. Safety, the prevention of accidental losses, is the finest investment any manager can make. It may not contribute to the maximization of profits, but, as Allen says, it will improve profits by minimizing accidental losses. In a number of companies, accidental losses have exceeded profits, and these organizations have gone bankrupt. Why? They did not put safety in the prime position.

3

Paradigm 2: Think Safe, Work Safe

MANY MAY HAVE SEEN OR HEARD these expressions or seen them posted at a work site. "Think safe." What does that mean? "Work safe." What does that mean? How many people have actually sat down and tried to analyze what is "think safe" and what is "work safe"? These expressions are meaningless.

As Bill Hoyle (1995) puts it, "'Safety First,' 'Think Safety,' 'Take Two,' 'Wear Your Safety Glasses' and 'Safety is Everyone's Responsibility.' These are a few of the many safety slogans found on banners and posters in most industrial facilities. What all these slogans have in common is that they send the message that accident prevention is primarily the responsibility of individual workers. When is the last time that you saw a safety banner directed at management such as 'Preventative Maintenance First,' 'Engineer It Safe' or 'Safety before Profits?'" (1).

Whenever I ask audiences, "What is safety?" they give me pages and pages of descriptions and quotes from dictionaries that define safety as *freedom from harm, freedom from injury, injury free,* or *accident free.* These warm, fuzzy expressions have no substance. Why think safe? Why work safe? These abstract concepts are too generic, too generalized, and they are, at the end of the day, meaningless. After all, why must a dictionary define safety when technical safety manuals have accurate, up-to-date definitions for the profession?

Let's ask the question again. Why must people "think safe" and "work safe"? Is it so that management can get on with the business activities and carry on running the show or so that management doesn't have to think safe or work safe? What do I mean by managers working safe? Are managers told by the board of directors and the chairman to "think safe" and "work safe"? No, because it is only employees who are told to "think safe" and "work safe." These are some of the thousands of meaningless, intangible, mottoes, and slogans used to try to engender safety change among the workforce.

Safety slogans and mottoes do little to change safety culture and habits. Thinking that they do is a paradigm. These slogans and chants form part of a worn-out safety mind-set. That safety expressions are meaningful and are a part of the science of loss prevention is a myth.

Managers often do what they want. They can drive around in luxury company cars without wearing safety belts and do inspections of the plant without wearing personal protective equipment. Perhaps they even go home and climb on their roof to shovel off the snow using a ladder that belongs on the scrap heap rather than a ladder that is safe. The workers at the coal face can't do what they want but are still the target of many senseless safety efforts in the form of safety slogans and mottoes, while often line management is exonerated from participation in the safety drive.

Reality

So what is reality, and what is safety? What is reality is to get the job done. I have spoken with numerous employees who are told, "Listen buddy, forget all this safety crap, we have a job to do, get it done no matter what!" This is safety in reality. Safety is seldom practiced first; getting the job done is what is first.

Getting the job done is not necessarily a bad thing. Employees are rewarded. They receive a pat on the back. They receive encouragement from their leaders when they get the job done. But should they be injured in this all-out rush to produce more, this is where the finger-pointing starts. Now, management may say, they did not "think safe or work safe." "You were injured," they accuse, yet it was insinuated that the job was the priority. The job was first. Only when there is an accidental injury does the employee who was getting the job done get blamed for not working and thinking safely.

At this stage, you are probably thinking, "But the employee could have refused to do a dangerous job." The truth and reality is that he *couldn't* have. How many employees would have turned to their superior and said, "Listen, this job is unsafe—I'm not doing it"? No one would do that without jeopardizing their job. This is a fact of life. When the supervisor says jump, employees say, "Yes, sir, how high?" Sure, the cop-out is that the employee can always refuse to do dangerous work—even the law says so. Who defines dangerous work? Does the supervisor? Does the employee? Or do they all agree after the accident that it was dangerous work. Or does the supervisor merely say, "Listen, don't be a wimp. Stop your whining. Do the job, or else. Nobody has ever been injured doing this job before." Sound familiar? Yes. As long as the employee gets all the tonnage that the supervisor requires, out and on time, things are fine. He can break all the rules because production comes first. Should there be an injury, then the employee will be blamed. After the accident, he will be told, "But you did not think safe or act safe!" Let the punishment and disciplinary procedures begin.

The following story from the *Daily Courier,* (1999) explains. The article tells of the owner of a Chino Valley sand and gravel open-pit mine that must pay $220,000 for safety violations in connection with the September death of an employee.

The U.S. Department of Labor's Mine Safety and Health Administration (MSHA) fined Paulden Sand and Rock for four violations, each with a civil penalty of $50,000, the maximum penalty amount MSHA allows, according to the press release.

The article continues that longtime Chino Valley resident Anton "Hank" Bisjak Jr., fifty-five, died when the front-end loader he was operating rolled off an embankment at the company's Perkins Pit. Bisjak, who had worked at the mine for two and a half months, was crushed and died immediately beneath the loader. He was married and had two grown daughters.

Investigators stated that the company knew the brakes were incapable of stopping and holding the loader and that the parking brake was defective. In addition, the company failed to provide a berm or guardrail where the work area drops off and failed to conduct a preshift inspection of the loader, which is a federal requirement.

Bisjak's father, Hank Bisjak Sr., a Chino Valley native, said that for weeks before the accident, his son compensated for the faulty brakes but never refused to operate the loader for fear of losing his job. "He used the transmission as brakes and it ran out of fluid," Bisjak Sr. said. "He said the company would give the job to someone else if he didn't do it. He couldn't afford not to."

Bisjak Jr.'s wife, Ann, said she is pleased with the investigators' ruling. "The government has done an excellent job of investigating," she said. "I think they're doing the right thing and I'm behind them 100 percent."

Dunbar Stone Company out of Ash Fork owns the mining company. Ann Bisjak said Dunbar is one of the largest sand and gravel companies in the Southwest, adding that the company hasn't compensated her family for her husband's death.

"The loader involved in the accident," Ann said, "is no longer in use after the crash totaled it."

The owner of the mine and the company's attorney were not available for comment when the article went to press.

In a Hurry

Have we, in safety, caught up to the thinking of the twenty-first century? While conducting a safety lecture, a trainee stood up and said to me, "You know, it is the twenty-first century, and people are in a hurry." That statement really opened my eyes because it is absolutely true. When we started pushing safety, people weren't in such a hurry. Today, time is money. Today, time is more money than it was yesterday. People are in a hurry. Communications are done in a hurry. Transactions are done in a hurry. Tasks that took forever years ago are now done in a few seconds. That's why the safety of the situation is often dictated by old-fashioned beliefs.

Look at the data processing and communication industry and see how things have changed. Time is of the essence. In safety, we have not kept up with the fact that people are doing things in a hurry.

Some years ago, my flight back from Alaska was delayed. The pilot was not happy with the brake warning indicators. The brake pads (or whatever they use on aircraft) were replaced within an hour. With only an hour's delay, the plane was once again made airworthy, and we flew off to our destination.

Twenty years ago, replacing the brakes on a similar aircraft would have been a four-hour operation. To do so, the plane would have had to be taken out of service and a substitute aircraft found to complete the journey. Being twenty years later, things change, and people are in a hurry. Artisans, laborers, and production workers are all under tremendous pressure to produce as much as they can within their shifts. No longer is there such a thing as a casual work shift. Stress is prominent. Profitability is the key word. Safety and the prevention of accidental losses have been pushed to the background. People are in a hurry. Automobiles are traveling farther and faster. Nothing much has been done to improve the driver's driving skills, but all the safety accessories, such as seat belts, air bags, and so on, are there to minimize the degree of injury should a collision occur. We are moving at such a pace that we have forgotten the importance of the skills of the driver. We have improved only the cushion, reducing contact around him once he is involved in a collision.

Because people are in such a hurry, safety has been left behind. The safety of human beings, the prevention of pain and suffering, and the reduction of risks have been put on the back burner because safety slows us down. Driving safely or doing a task safely hinders our performance and acts as an anchor to our fast-moving pace. Safety is no longer first. Safety hampers progress. Safety reduces throughput. Safety forces people to take their time. Safety is often the last consideration in this fast-moving society.

Intangible Concept

Thirty years ago, I was sitting in my first safety training session. The managing director of the National Occupational Safety Association, H. J. Matthysen, was leading it. He conducted a four-hour training session, and I remember his one statement as clear as if he made it yesterday: "Safety is an intangible concept. You cannot see it, you cannot feel it, you cannot touch it, and you cannot hear it. The role of the safety professional is to convert this intangible concept to tangible concepts."

Today, some thirty years after I learned this lesson, safety is still very much an intangible concept. Few people really understand the concept of safety. Most people think that safety has to do with "freedom from injury." Most people think that safety belongs with some other organization or department. A number of people, including management, believe that safety is what "safety people" do. These are part of safety's paradigms.

A dynamic speaker once taught me that the "KISS" rule applied. By the KISS rule, he meant, "Keep It Straightforward and Simple." This speaker kept his presentation straightforward and simple. I remember the impact that he had on me and other members of the audience. We have failed to keep safety straightforward and simple. The KISS in safety does not exist.

Some of the greatest personal successes I have had in the safety profession have been taking the components of safety and converting these to tangible, achievable objectives. This makes safety tangible. So does the KISS principle.

The one-day coaching course that we introduced as part of our training program did just that. This coaching course took the intangible concept of safety and converted it to seventy tangible concepts that, when combined, formed a safety system that epitomized safety per se.

"Think safe, work safe" are intangible concepts. It is a ridiculous, untruthful concept and as such cannot be understood and cannot be conveyed to the workforce or even explained.

Tangible Concepts

My point is that if we want to sell safety, we must break it down into tangible concepts. Once it is broken down to tangible concepts that people understand and can do something about, we will then be putting safety in a recognizable form. It will not necessarily become an immediate priority, but it will be in a position to be reckoned with *once it can be recognized.* The only way to recognize safety is to convert it from these intangible, warm, fuzzy concepts of "think safe," "work safe," "safety is a priority," and "safety is first" to the basics, such as housekeeping, machine guarding, demarcation of walkways, training, risk assessment, and correct procedures.

My success in the safety field has been due to the fact that I simplify safety. I show people what safety is. I escort people through their own workplaces and point out items such as machine guarding, electrical safety, and the wearing of personal protective equipment. People can relate to these safety elements. Put enough critical safety elements together, and you have a safety system. People can now understand what a safety system is because they understand the components. Nobody understands the components of "think safe, work safe." That is why the safety message often fails to get through. Understanding, recognizing, and appreciating the tangible concepts of safety will start getting safety recognized as a competitor in the game. Presently, safety has lost its poll position, even though people still chant, "Think safe, work safe."

Using worn-out, meaningless safety mottoes and expressions will not achieve the goal of risk reduction to eliminate accidental losses, and thinking that their use is going to fix the safety problem is one of safety's paradigms.

Safety Posters

A "safe" plant dresses itself up in colorful safety posters, and this prevents accidents and makes the workers careful workers. Safety posters are another old-fashioned safety tactic and contribute little to safety awareness.

During thousands of safety audits, I have often questioned workers on the message of the posters on the workplace wall. Most reply to my question, "What poster, what wall?" Safety must be promoted by more modern and effective methods, not by outdated methods such as posting safety posters on the walls. Thinking that they contribute to the identification of and mitigation of workplace risk and at-risk behavior is another safety paradigm.

4

Paradigm 3: No Blood, No Foul

THE "NO BLOOD, NO FOUL" CONCEPT means that the workplace is safe until there is an accident and a resultant injury. A common paradigm is that if there is no injury, there is no accident. As long as an organization goes from day-to-day injury free, then safety is fine. Safety is first, yes; nobody is being injured, so things are in order. Meanwhile, the organization could be experiencing numerous undesired events (accidents). To explain, an accident is an undesired event. An accident is an undesired event that results in some form of loss. This loss could be in the form of injuries to people, damage to property or equipment, or merely business interruption or other things that hamper the input, throughput, and output process. All these "things," or undesired events, as they will be referred to, are accidents. They are unplanned events. They need not end up causing injury. As unplanned events, they therefore qualify as deviations from the management planning.

Most seasoned and experienced safety professionals will agree that judgments of safety performance based solely on injury frequency rates are grossly incorrect and inaccurate. If one gauges a breakdown in the management system by counting injury losses only, a false impression is created. Therefore, a company that is ticking over, injury free, is not necessarily safe. In fact, it *could* be experiencing near misses or events that, under slightly different circumstances, *could have* resulted in injury. It is definitely experiencing undesired events that are hampering the process. Equipment is being damaged. Products manufactured are being damaged and become, as is often the case, unsafe or below quality standard products and are exported around the world.

So, if a business is injury free, it is not necessarily safe. Behind the scenes, the tip of the iceberg may not yet have appeared, but the base exists. A safe organization means no deviations in the management system. No poor-quality products are produced. Production quotas are met without losing money in other areas, such as injuries, damage, and interruptions.

The Health and Safety Executive of Great Britain (2006) stated, "Any simple measurement of performance in terms of accident frequency or incidence rates is not seen as a reliable guide to the safety performance of an undertaking." They suggest that the efforts to control foreseeable risk be assessed instead.

Because there is no blood does not necessarily mean there is no foul. Think this through. Every single undesired event has the *potential* to produce injury. Are the undesired events being prevented? Is the prevention of deviation from management standards a priority in your business? If not, numerous fouls are taking place, but because there is no blood, we do not recognize the foul. The point is that because an organization is not injuring people does not mean that it does not have a safety problem. Safety is defined as the *control of all accidental losses.* Is the organization controlling *all* accidental losses or merely suffering from the "ostrich syndrome"? The ostrich, by the way, is indigenous to South Africa. When an ostrich gets flustered, stressed, or threatened, it sticks its head into the sand, literally, and ignores all the problems around it. It thinks that the danger has gone away. No blood, no foul is a clear example of an ostrich sticking its head into the sand. Absence of injuries is no indication that the safety is in order.

Why Injuries Are Seen as Safety

One of the main reasons for the paradigm of no blood, no foul is that safety is perceived as being about injury to people. This is another mind-set that hampers safety efforts. Only when somebody gets injured does safety become important. It becomes the talk of the day for a few hours. People's minds heal quicker than the injury. Workers will mill around, drinking coffee, discussing injury to a fellow worker or other employee. The next shift, it is business as usual. Joe's okay, he is recuperating in the local hospital. Let's continue, it'll never happen to me. What about all the near misses, close calls, and other no-loss-producing events that, under slightly different circumstances, could have caused injury?

Damage

Few, if any, regard damage to equipment, products, and goods as being the result of an accident. What about poor quality? Is that as a result of an accident?

The answer is yes. If you regard safety as the control of accidental losses, then product recall is certainly a safety issue.

Property damage accidents exceed the number of injury-producing accidents. According to Bird and Germain (1996), for every serious injury there are at least thirty accidents that result in property damage. During their 1966 study of nearly 2 million accidents reported to the Insurance Company of North America, they found that thirty property damage accidents were reported for every major injury. They also stated, "Property damage accidents cost billions of dollars annually and yet they are frequently misnamed and referred to as *near accidents*" (21).

Their line of thinking recognizes the fact that each property damage situation *could have* resulted in personal injury. This injury mind-set is a holdover from earlier training and misconceptions that led supervisors to relate the term "accident" only to injury.

In reviewing the Bird–Germain ratio, only 1.6 percent of all accidents occurring at a factory or mine result in injury. Only 0.15 percent of accidents (undesired events) cause *serious* injuries, which are known as *reportable, lost time,* or *disabling injuries.*

Why this category of injury is used is because this is the *degree,* or type, of injury that is mostly recognized in the safety profession. This injury is the only thing that safety campaigns, safety records, safety performance, safety comparisons, safety competitions, and safety recognition schemes recognize as being the *numero uno* of safety. If less than 1 percent of serious injuries are caused by accidents, why do most safety efforts concentrate on them? What about the minor injury-causing accidents? What about accidents that cause property damage? What about the numerous near misses? All the previously mentioned have *potential* to result in severe injury, if it were not for the three luck factors that determine the course of the undesired event.

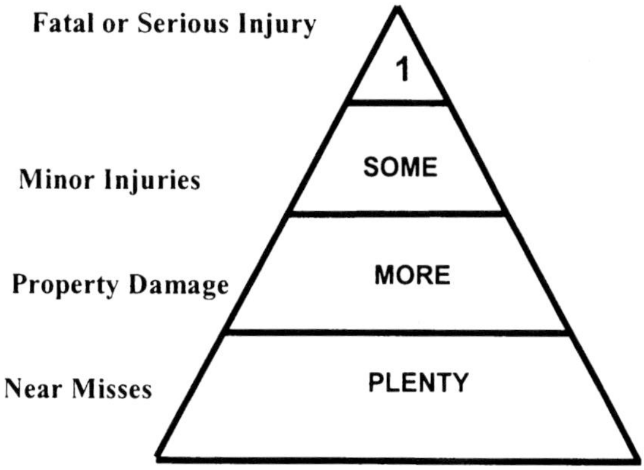

Model 1. The principle behind the accident ratio.

Accident Ratios

Numerous professionals have proposed accident ratios, which give the ratio of serious injuries to property damage accidents and near misses. H. W. Heinrich proposed the 1:29:300 accident ratio as far back as 1929 in his book *Industrial Accident Prevention.* What Heinrich (1959) stated was that for every 330 undesired events, only one resulted in serious injury. Bird and Germain said exactly the same thing after their 1966 accident ratio study. They proposed that for every 641 undesired events, or deviations from management plan, only one would result in serious injury. The Health and Safety Executive of Great Britain proposed something similar. Safety pioneers around the world have derived other ratios. The facts remain: there are numerous near misses (warnings) before a serious injury-causing accident is experienced. Changing the paradigm that if there is no blood there is now foul is to heed these warnings before it is too late.

The fact remains that for every *one* serious or disabling injury, there are *some* minor injuries, *more* property damage accidents, and *plenty* of incidents where nothing happened but where something *might have happened.*

The bottom line is that all undesired events fall under the heading of *safety.* In professional safety, all undesired events should be examined. The risk of these events should be assessed, and those with the highest probability, severity, and frequency should be investigated. Once the basic causes are sought, remedial measures should be put into action to prevent this deviation from standards recurring. Only by preventing a deviation occurring will the consequence—or the likely consequence—be eliminated. Safety is often reactive because we sit back waiting for the consequences to occur before we take action. This is reactive safety.

Iceberg Effect

The tip of the iceberg is what is visible. The tip of the safety iceberg, however, very seldom does the most harm. It is the base of the iceberg, or the larger, hidden portion, that creates the havoc. The portion of the iceberg *under* the waterline is what sank the *Titanic.*

Minor injuries, property damage, and near misses constitute the hidden part of the iceberg and are the parts of the undesired events that can wreak havoc with a business system. Safety efforts cannot be successful if they focus on the tip of the iceberg. Safety, if viewed as the control of total accidental losses, must focus on the base of the iceberg. The bottom portion of the iceberg represents the accidents that a company has not yet had. These are near misses, or warnings that consequences could follow, *under slightly different circumstances.* Remember that luck determines the outcome and end result of an undesired event.

Incident Recall Case Study

On conducting a safety presentation one morning, I arrived at the lecture venue early. On entering the room, I noticed that the lights were turned off and that the area was in darkness. Looking up, I immediately saw why. An electrician's assistant was standing on the very top of a self-standing stepladder and was removing the fluorescent tubes from the light fitting. I assumed that he was busy with the replacement of the tubes. I returned to my car to collect another load of books and course handouts.

As I reentered the training room, carrying a box full of books, the training instructor walked in behind me and said, "Why don't you turn the lights on?" and with that he threw the light switch. The entire training room suddenly lit up with all the lights having been turned on to reveal a startled and terrified electrician's assistant with his hands full of electrical wires from the workings of the fluorescent lamp now connected to the main power supply. I'll never forget the look on his face. His eyes were wide open, and he was frozen, almost as if he had been turned momentarily into a statue.

In that split second, it became obvious that he had received the fright of his life as he knew he was working with current-carrying conductors, and he had failed to isolate the circuit breaker, lock it out, or tag it.

Although this event could have ended up in the electrocution of the assistant, no action or investigation whatsoever came about as a result of the near miss. The luck factor here was that the event did not result in any serious consequent or loss, so "nothing happened": no blood, no foul.

Near Misses Aren't Important: A Paradigm

The ignoring of near misses or warnings as not significant is a major safety paradigm. For safety to be effective and to take its correct place next to production and quality, near misses must be recognized and acted on. Near misses are defined for the purpose of this book as *incidents.* The definition of an *incident* is *an undesired event that, under slightly different circumstances, could have caused injury to people, damage to property, business interruption, or harm to the environment.* These have often been called the accidents that have not yet happened. Most organizations are years behind with the recognition of the importance of incidents. Most organizations are just as far behind when it comes to the reporting, evaluation, and rectification of these warning signs.

Nothing Happened

The apathy that exists in organizations, in terms of recognizing near misses, is the false sense of security that "nothing happened." Managers are lulled into a false sense of security because nothing apparently happens. Employees don't report these events because "nothing happened." Again, it's a matter of no blood, no foul.

Near misses constitute the majority of undesired events occurring in workplaces. The only difference between a near miss and a hit (which causes property damage or injury) is luck. In fact, three luck factors determine how the dominoes are going to fall and what the end result will be.

It's Not What Happened . . .

The most important sentence in the entire philosophy of safety management is, It's not what happened, but what *could have* happened. This means that each near miss could have resulted in something else if it were not for luck. Almost all individuals in every group I have trained have admitted that they have experienced near misses. Near misses are understood by hourly employees, middle management, the safety profession, and the executive management. If only some action would be taken on near misses, the root of the safety problem would be tackled.

Injury-Free Culture

I have often heard the expression "injury-free culture" and other expressions just as warm and fuzzy. Some of them sound like this: "We are striving for an accident-free culture." Others boast, "We want to be injury free for a year," and perhaps the biggest of them all, as far as safety is concerned, is, "We will have zero accidents." Even though it sounds good, it is almost impossible without bending the rules and hiding the mistakes. Many organizations create an aura or climate of fear so that terrified employees do not report accidents, near misses, or injuries, and the company fools itself that it has achieved its injury-free goal. I hate to burst the bubble of the authors and users of these expressions, but the following truths must be recognized:

- You cannot be injury free unless you are accident free.
- You cannot be accident free unless you are incident free.
- You cannot be incident free unless you have no unsafe conditions and no unsafe acts are committed.
- You cannot eliminate unsafe acts and conditions until you have effective control measures in the form of a structured safety system.
- Your safety system cannot be effective until you assess and reduce the risks.

So, the bottom of the iceberg is what sank the *Titanic.* The bottom of the safety iceberg represents the numerous near misses that occur before there is an accident that causes property damage or minor or serious injury. Focusing on the top of the iceberg, namely, the injury, is futile. You cannot eliminate the serious injury at the top of the safety iceberg until the near misses have been reduced substantially. Reducing the number of near misses reduces the probability and frequency of an injury occurring.

Reporting Near Misses

Near misses cannot be assessed or managed until they are reported. The finest mechanism for reporting near misses is the workforce. The workforce must be encouraged and enthused to report near misses and should not be penalized when they do so. Some of the main obstacles to the reporting of near misses are as follows:

- The employee is held accountable for the near miss.
- The employee is challenged with "what have you done about it?"
- Nothing gets done.
- The employee is punished.
- Employees see no action on the report.
- Employees do not receive recognition for reporting.

I have worked with organizations that receive between 700 and 800 near miss reports every month. These near misses are reported by hourly employees, middle management, and senior management. Everybody is involved in the near miss reporting system. What makes it even better is the fact that each near miss that had been reported had already gone through a risk assessment process by the person reporting. The trick to a near miss reporting system is to make it a "no names" reporting system. Discipline must be entirely removed from the system, and employees must feel comfortable to report these undesired events. If an incident, with the potential to kill somebody, is reported, no punitive action should be taken. This will deter people from reporting near misses. To make safety work, we should play the ball and not the man.

Near Misses Left Behind

I once presented a lecture at a regional safety congress in the United States and asked the audience how many near misses they had reported at their company during the preceding year. Of an audience of seventy people representing major organizations, only three indicated that their organizations had reported near misses. This led me to believe that most organizations were left behind as far as their safety systems were concerned.

The reporting and ranking of near misses is perhaps the most important aspect of any safety program. The near misses with high probability and high potential severity should get the same important treatment as actual accidents. They warrant a full accident investigation, complete remedial measures, and follow-up.

Instituting a structured near miss reporting and rectification system can make the biggest advancement in safety in any organization. I have personally read the monthly printouts of excellent near miss reporting systems. Seldom does a month go by without at least two of the near misses having the potential to have caused fatal injury. This means there is an opportunity to rectify the causes of these near misses before the consequences are experienced.

Warnings in Advance

The advantage of the safety profession is that it is the only profession in the world that gives warnings of imminent danger. What I mean by that is, according to Frank E. Bird Jr. and George Germain, there are 600 warnings before a serious injury occurs. This is derived from the Bird–Germain ratio, which has been discussed.

I must emphasize here that the only difference between a near miss and a hit (exchange of energy) is luck. Numerous writers have come to the same conclusion. These include Frank E. Bird Jr., H. W. Heinrich, William E. Tarrants, Dan Petersen, and others. Yet the safety profession seems to ignore this valuable information. Safety can be successful only if we expose the paradigms for what they are and if the near misses are identified, reported, and acted on.

Luck Factors

If the organization has a low injury rate, one must ask the question, "Is it as a result of *control*, or is it as a result of relying on *luck*?" If it is as a result of relying on luck, then I suggest the safety system be reviewed. An organization must make sure that it has at least seventy elements that constitute its safety system that puts it in control of the things that could lead to undesired events.

The luck factor is so prominent and yet so poorly understood. Even working a million man-hours without a disabling injury could be as a result of luck. Not having a fatality for the past fifty years could be as a result of luck. Having a fatality once a year could be bad luck. The point is, unless the controls are in place, we are relying on luck, not control and risk reduction, which prevent the events occurring in the first place.

Examples

During a safety lecture, a student challenged me as to what I could teach them about safety. When I asked him what he had meant by his statement, he confronted me and said, "You fellas in South Africa have just killed 105 miners in an underground accident. You dropped a twelve-ton locomotive on a moving cage, and this forced the cage to the bottom of the shaft, where it was crushed, killing 105 people. What can you teach us about safety?"

I was rather taken aback by this accusation, even though this young man was absolutely correct. On May 10, 1995, at the Vaal Reefs No. 2 Shaft in South Africa, a locomotive pulling a cage weighing thirteen tons in total went past the safety stops at the 56 Level and plunged down the shaft. The locomotive and cage crashed onto a moving double-deck cage, about 150 feet below the level, crushing the cage's top deck, snapping the cable, and forcing it to plunge 1,500 yards to the shaft bottom, where the cage was crushed from a height of twenty-one feet to three feet eight inches. The 105 occupants of the cage suffered a horrendous death.

The young man was fully within his rights to tackle the integrity of South African mining safety. I was speechless for a while because the Vaal Reefs disaster had sent the entire South African nation into mourning. The then president of South Africa, Nelson Mandela, made an appeal on national television to the mining industry to return the spotlight to the safety issues confronting the deep-level gold mining industry. There were upheavals in the ranks of the mine union and conflicts between mine management and unions. The then minister of mines and mineral affairs, Pik Botha, said after the accident, "Surely it wasn't so difficult to have prevented it happening all together."

As this particular lecture happened to be at an underground mine, I then asked this young man if they had ever dropped a locomotive down one of *their* mineshafts. There was a deadly silence in the room. Then an old-timer who had been mining there for thirty years put up his hand and said, "Yes, in 1960 we dropped a locomotive down No. 3 Shaft." I asked him how many people had died, and he replied, "None, no one was even injured." I could feel the tension grow in the room as I asked the old-timer, "Why? You dropped a locomotive down the shaft the same as Vaal Reefs, yet nobody was hurt, and nobody got killed. Why?" The old-timer looked around at the other students in the room and then looked at me and said, "I guess we were lucky."

The point being made here is the fact that the difference between this particular organization's locomotive falling down a shaft and the Vaal Reefs disaster was a matter of luck. Vaal Reefs was unlucky that the cage was *below* the level from which the locomotive plunged. The mine that I was at was fortunate that the cage was *above* the level from which the locomotive plunged down the shaft so that the locomotive fell to the bottom of the shaft without colliding with the cage.

The luck factor is very prominent in determining what is going to happen. Near misses are indications of what *could* happen. Put them together, and one can almost predict the next severe accident.

William E. Tarrants (1980) reaffirms the importance of the luck factor: "The general justification for the collection of non-injurious accident data is that the severity of accidents appears to be largely fortuitous. Accidents from the same cause can recur with high frequency without a result in injury" (46). He agrees that accidents that do produce an injury appear to be determined largely by chance.

Confirming the importance of risk ranking of near misses, Tarrants also said that it is now possible to identify and examine accident problems "before the fact" instead of "after the fact" in terms of their injury-producing or property-damaging consequences. He stated that this allows the safety professional to concentrate on the measurement of *loss potential* of near misses and removes the necessity of relying on measurement techniques based on the probabilistic, fortuitous, rare-event injurious accident.

Risk Assessment

In referring to the fifteen-domino *cause, effect, and control of accidental loss* sequence (McKinnon 2000), the first domino that depicts the beginning of the entire chain reaction, called an accident, is the "failure to assess the risk." Risk assessment has been so confused and misunderstood by safety professionals that it warrants time here to simplify how to do a risk assessment.

A basic risk assessment is merely the asking of two questions:

1. What is the possibility of its happening?
2. If it happens, how bad can it be?

Once a hazard has been identified, ask, "What can this hazard cause?" In other words, "What are the chances of it happening?" and also, "How bad can it be?" Both these questions are asked with the normal expected loss in mind. Worst-case scenarios are not used for basic, simple risk assessments.

The ranking of these two scenarios could be a simple ranking such as *low*, *medium*, *medium-high*, or *high*. In other words, the *probability* of somebody being injured by this accident is, for example, *high* or *medium-high*. For the *loss potential*, one could rank it as *low*, *medium*, *medium-high*, or *high*. High loss potential could be a serious injury, and the low could depict minor injury or minor interruption.

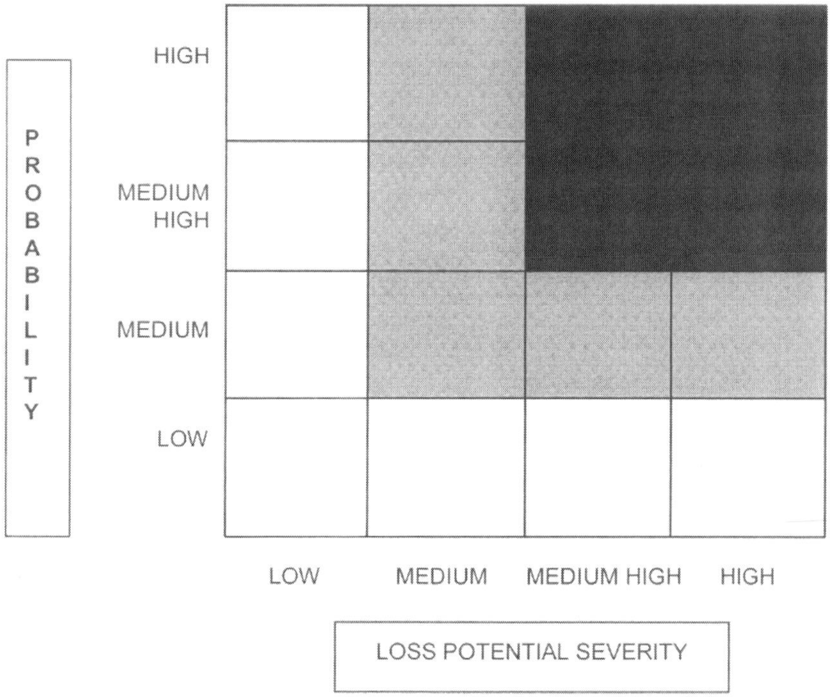

Model 2. An example of a risk matrix. X-axis refers to the loss potential severity, Y-axis refers to the probability of occurrence.

All the near misses in the *medium-high* or *high* category and higher should receive proper accident investigations.

To assist you in convincing others of the importance of near miss reporting, a thumbnail sketch of a near miss and near miss reporting is given under the following heading.

Near Misses and Near Miss Reporting

The history of thousands of undesired events has shown that the outcome of the event cannot be predicted and that under slightly different circumstances the consequences could have been worse if it were not for the luck factor.

Unsafe acts and conditions are called the *immediate* causes of accidents. They are the obvious causes or the causes that lead to the contact with a source(s) of energy. The principle of multiple causes indicates that accidents are usually the result of multiple causes.

Near misses have often been referred to as "safety in the shadows," as this is where the heart of the problem lies. Near misses offer management an opportunity to rectify a system breakdown before it happens. Because there are no losses as a result of an undesired event does not necessarily mean that the event is insignificant. Risk assessment of all near misses will determine which will warrant a full investigation to track and eliminate the source of the problem at the root.

Incidents (refer to the definition given previously) are those near misses that were closest to causing some form of loss, as there was an actual exchange of energy. The exchange was insufficient to cause loss or injury, but the fact that there was an exchange of energy is reason enough to heed the warning. Remember, it's not what happened but what *could* happen.

Each accident results in some form of loss, and all losses cost money. Time may be lost, forms need to be filled out, and the business is interrupted to a degree. Many of the costs of an accident are *hidden* and therefore go unnoticed. *Direct costs* or insured costs are normally the only costs associated with an accident. They are the lesser of the two costs. A third, deeply hidden layer of costs is the *totally hidden costs*, which are seldom identified or tallied.

Because an injury is minor does not mean that the event that caused the injury was minor. The event should be investigated and the potential and probability of recurrence evaluated. The next, similar event may have far worse consequences because of the three luck factors (under slightly different circumstances).

Measure Safety

Most safety programs count the serious injuries as a measure of "safety." This measurement method, while still accepted, is in fact a measure of failure. Auditing the activities that make up the control measures is what should be done to measure safety. This result is a more positive measurement of management work being done to combat loss. This is what "safety" is.

One cannot focus on the tip of the iceberg, as the tip is the result of the base, which, in the accident ratio, is the unsafe conditions and behavior that led to the losses. Focusing on the tip (the serious injuries) is treating the symptoms of the problem and not the cause.

Your greatest safety tool is the risk matrix. Remember, it's not what happened but what *could* happen. The risk matrix is a crystal ball to predict the future or possible outcomes of an event. Use it to forecast the probability of the next loss.

High-loss-potential events that did not result in loss should be investigated as rigorously as serious injury-producing accidents if the assessment of the risk shows in the black on the matrix. Risks that rank in the gray area should also be subject to an investigation. All injuries and loss-producing events (in excess of $1,000) should always be investigated.

Reporting all undesired events such as near misses and incidents is perhaps the most important aspect of any safety program. A no-blame system should be introduced to encourage reporting without consequence. Most near miss programs fail as a result of disciplinary steps being taken once an event has been reported. The more warnings are turned in, the more the opportunity to investigate, identify, and rectify the root causes.

Summary

Near misses have often been referred to as "safety in the shadows." Near misses (incidents) occur far more often than loss-producing accidents do. They are warnings and should be heeded. According to Frank Bird Jr., for every serious injury, there were 600 other occasions that could have led to the injury. Safety can take a quantum leap by identifying, risk ranking, and rectifying those incidents with high loss potential.

Workplaces and homes will never, ever, be injury free. No matter what degree of safety control and effort is put in, all accidents will never be eliminated. Safety programs, safety campaigns, and safety systems cannot eliminate injuries. Injuries—or the resultant losses—are determined by luck factors.

What all these proactive efforts do (and that is *all* they do) is reduce the probability and the frequency of the event occurring. You cannot entirely eliminate the event from occurring and causing injury or property damage. You can make the efforts to reduce the *chances* of its happening. If it happens, remember that luck determines the outcome anyway.

5

Paradigm 4: There Are No Cop-Outs in Safety

SAFETY HAS BECOME THE HOT ITEM that most want to divorce themselves from and not get involved in. This has led to a situation where employees are normally blamed if an accident occurs, property is damaged, or somebody is injured. The roots of this situation go back to H. W. Heinrich's theories and findings, which he published in his book *Industrial Accident Prevention: A Scientific Approach* (1959). In this edition, he stated that research that he had done, using the most prominent accident causes, showed that 88 percent of accidents were caused by unsafe acts of people. He stated that the unsafe work environment caused the other 10 percent of accidents. The final 2 percent were, according to him, caused by acts of providence or as a result of natural factors or things and items over which we have no control.

His ten axioms on industrial safety are well known throughout the safety world. His second axiom states, "The unsafe acts of persons are responsible for a majority of accidents" (Heinrich 1959, 21).

Statement

This is the statement that set the course for safety professionals many years ago and is one of safety's major paradigms, as his research was flawed. Maybe this was applicable at the time of Heinrich's research, but this statement was based on poor accident investigation reports, and only the most significant contributing cause of accidents was taken into account in compiling these statistics. Who decided what the most significant cause was?

It is obvious that the level of accident investigation seventy years ago was poor and that employees were found to be at fault and the cause of accidents. The unsafe act would therefore feature prominently as a cause in accident investigations. Heinrich's domino sequence on loss causation also indicated that one's ancestry and the social environment in which one was raised created the propensity for people to make mistakes (fault of person), which caused the unsafe act and unsafe condition. In the five-domino loss causation sequence, two of the dominoes, the first and the second, pointed to the fault of people.

Unfortunately, this statement is still believed, taught, and accepted today. It is obvious that this statement is no longer applicable. Having conducted my own research over a three-year period in a large industrial undertaking, I discovered that in one year some 63 percent of accidents were caused by unsafe physical conditions. The other years gave slightly different results, but I also used only *the most significant accident cause.* Statistics can be manipulated. Seminar leaders who make the opening statement that 97 percent of accidents are caused by the "at-risk behavior of employees" are really taking a wild guess, jumping to conclusions, and proposing absolute assumptions. They create the paradigms, and we believe them.

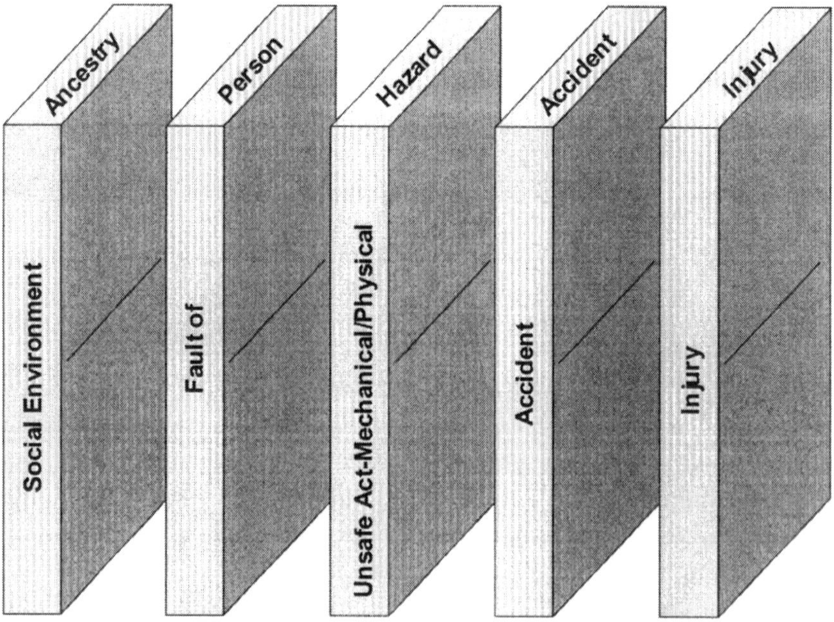

Model 3. Heinrich's domino theory.

Bill Hoyle said that behavior modification safety programs are a perfect fit for management's avoidance of responsibility for health, safety, and accidents. When these programs refer to unsafe behaviors, they are referring to worker behavior only. He continues by saying that the decisions of corporate chief executive officers, presidents, managers, engineers, and supervisors are the primary factors in preventing accidents. When management runs process units beyond factory design limits, reduces preventive maintenance staff, crowds equipment together, hires poorly trained contractors, and downsizes safety budgets, these unsafe management behaviors are ignored by behaviorist safety programs.

Too Simplistic

By blaming employees for accidents, we are really oversimplifying the safety problem. By making statements such as "The majority of accidents are caused by unsafe acts of people," we are diminishing the importance of a safe work environment. We are also watering down the impact that good management has on the reduction of accidental losses. This is one of safety's sacred cows, and few have the heart to challenge it.

Many people say that a person must have created an unsafe condition anyway. Because of this, they say, "Well, all hazards have to do with people; therefore, people are responsible for the majority of accidents." This is far too simplistic. The actual causes of accidents within an organization can be determined only by a careful study of all the facts and data and by compiling statistics applicable and site specific to that organization and by uncovering the root causes.

It's a Cop-Out

By pointing a finger at employees and saying that unsafe acts of people cause the majority of accidents, we are creating a cop-out situation. This is a good excuse for line management to maintain below-standard physical and mechanical work conditions and to pass the buck entirely to employees. To preserve the safety record, organizations blame employees. The four-finger rule entails pointing a finger at an employee and stating, "You committed an unsafe act!" Meanwhile, the other three fingers on the hand point directly back to management, asking, "Were you not partially responsible?" This is certainly a tough question to ask, and numerous managers, on various levels within different organizations, have never, ever, asked the question. Perhaps they were afraid to. The three fingers pointing back will ask the following questions, among others:

- Have we trained this man adequately?
- Have we motivated this person to work safely?
- Have we created a work environment that is conducive to safe work practices?
- What are our system standards and controls?
- When did we last audit our safety controls?

If a change is ever to be made in the safety management profession, this is it; we must stop assuming that employees decide whether they want to work safe. We should disregard these one-sided statements and myths such as "Unsafe acts cause the majority of accidents." These statements have not been proved, and until they are, we should abstain from using them. Do internal studies and determine what are the immediate causes and the basic causes of accidents in your plant. By doing a root-cause analysis of every undesired event, you will open a can of worms and uncover the truth.

Multiple Causes

The principle of multiple causes, as applied to safety, simply states, "Accidents are seldom, if ever, the result of a single cause." What this means, in plain language, is that there is always more than one cause for every accident or near miss. By focusing on the unsafe behavior of people only, we are finding fault, not the facts.

The principle of multiple causes applies to the immediate causes of the accident, the unsafe acts, and unsafe conditions. The principle also applies to the basic, or root, causes of the accident. These are normally defined as the personal and job factors.

I recall having a conversation with a miner who had worked underground for forty-two years. He told me that he was completing an accident investigation form one day and was having a discussion about the unsafe condition that existed when his supervisor shouted to him from a distance, "There's no such thing as an unsafe condition in this mine!" They were forbidden to use the term "unsafe condition" on the accident investigation form. So management goes scot-free, and the finger once again points to the employee.

Mark A. Friend

Friend (1997) discusses certain statements concerning controversial safety issues. He says that although 90 percent of all accidents and adverse incidents may in fact be caused by human error, safety efforts must also encompass engineering principles. When asked whether the engineering or human relations school should be emphasized, Fred Manuele (1997) says, "A curse on both their houses" (31). Friend states that a seminar leader once suggested that since 95 percent of all safety problems are due to human error, 100 percent of the solutions should address the human element. His reasoning is that by taking such action, nearly all problems will be covered. Such thinking is naive and dangerous, according to Friend. Some problems can be solved, by addressing the human element. But Friend warns that ignoring engineering solutions is akin to fixing the problem with a short-lived solution that will fail when the slightest change occurs.

H. W. Heinrich

H. W. Heinrich was one of the first safety pioneers to put safety into the context of scientific management. His research is to be praised, and many of his principles are still taught today. However, taking his research and applying it in the twenty-first century requires that the research be redone in modern industry.

Hundreds and thousands of safety professionals have quoted Heinrich's axioms and the 88:10:2 ratios. The question is, did they read the chapter completely? If they had, they would quote those statistics and explain how they were derived. They would indicate how misleading and inaccurate they were because of the fact that the researchers considered only the most *significant* cause of the accident. Obviously, since then, somebody saw a chance of pinning all accidental losses on the fault of employees and used this research to their advantage. Heinrich's domino accident sequence was used and taught until the late 1960s, when Frank E. Bird Jr. and George Germain produced the updated Bird accident sequence.

Frank E. Bird Jr.

The Bird accident sequence revolutionized Heinrich's thinking in that Bird suggested that the first domino, or initiating event, in the accidental loss causation situation was poor management control. This poor control then triggered basic causes in the form of personal and job factors. These basic causes led to the unsafe acts and/or unsafe conditions. These immediate causes, as the acts and conditions are known, led to the contact with the source of energy and the resultant loss. Bird also included all aspects of accidental loss and not only personal injury.

Bird versus Heinrich

Even thirty years after being published, the Bird domino sequence still seems to take a back seat to Heinrich's original works, first published in 1929. I asked myself why? Bird, a professional safety practitioner, with nearly forty years of experience, conducted the most extensive safety research ever done to compile his loss causation domino sequence. Yet why is his cause and effect of accidents not being used as widely as Heinrich's? In fact, modern certification programs around the world still teach Heinrich in preference to Bird.

I believe that Heinrich is quoted in preference to Bird because Heinrich's principle suits organizations better. Heinrich, incorrectly quoted or simplified, states that people are responsible for their own safety, that the unsafe acts of people cause the majority of accidents, and these acts are caused by their "ancestral background," and the "faults of persons."

Bird (1992), to the contrary, states, "An accident is caused by poor management control" (30). He goes further and says that safety is *management* control of accidental loss.

These sophisticated, well-researched statements are nevertheless still preempted by the Heinrich philosophy, as it is easier to buy in to a cop-out than to accept the fact that management has any responsibility in safety whatsoever.

Research

To identify and analyze the causes of injury and property damage accidents at a large manufacturing plant, statistics were analyzed for 1996, 1997, and 1998. As with Heinrich, a decision was taken to list only the *prime* cause of the accident.

The following analyses were found:

- During 1996, 352 undesired events that resulted in loss were recorded. Two hundred and eight were a result of unsafe acts and 131 a result of unsafe conditions. This meant that 60 percent of the accidents were a result of the unsafe acts and the balance, 38 percent, a result of the unsafe conditions. There were some borderline cases that were not included for this analysis.

- In 1997, the same research was conducted under the same conditions, and 388 loss-producing events were recorded. One hundred and thirty were a result of the unsafe acts and 180 a result of the unsafe conditions. This calculated to 33 percent of the accidents being caused by the unsafe acts and the majority, 46 percent, by the unsafe physical conditions.
- In 1998, the same research was conducted at the same plant. Three hundred and forty-one loss-producing events were recorded. One hundred and seven were a result of the unsafe acts and 160 a result of the unsafe conditions. This meant that some 47 percent of all accidents were a result of the unsafe conditions and the minority, or 32 percent, the result of unsafe acts.

This plant could therefore quote actual, real statistics as to the cause of their accidents. The flaw in these results is the same as in Heinrich's research: only the most significant cause was used to calculate the results.

However, because all accidents have multiple causes, one cannot single out one cause, add them together, and produce an acceptable statistic. Multiple causes must be considered. Even if a cause is a basic, almost hidden one, it may have contributed to the unsafe act or condition that led to the contact and subsequent loss.

DuPont

DuPont has a worldwide reputation as being the safest organization in any country. To qualify how safety is measured in this context, it is measured solely by the number or absence of reportable or disabling injuries. In the National Safety Council publication *Accident Facts* (1997), DuPont is quoted as having the longest-running record of injury-free man-hours, which at the time of quoting was 66,645,000 man-hours.

During their safety seminars, DuPont are adamant that their research has shown that 97 percent of injury-producing accidents are caused by the unsafe acts of people. That may be particular to their type of industry but should not necessarily be taken at face value as being relevant to other industries.

Their statement concerning unsafe acts is too generalized, as true safety involves far more than counting the number of disabling injuries. Disabling, lost-time, and reportable injuries are *degrees* of injury, dependent largely on three luck factors and, of course, on honesty of reporting. The definitions and qualifying criteria of injuries are also highly subjective, as will be discussed.

The following article from *Occupational Hazards* (1999a) shows that hiding injuries is apparently common.

> DuPont Fined for Recordkeeping
>
> E. I. DuPont De Nemours & Co., failed to record 117 occupational injury and illness cases during 1997 and 1998, and recorded other cases incorrectly at its Seaford, DE plant, according to OSHA.
>
> At the time DuPont were to have paid a $70,000 fine and planned to implement a series of employee health and safety improvements under an agreement with OSHA, which issued the company one unclassified citation and two other-than-serious citations. The citations resulted from an OSHA inspection, which was initiated in response to a complaint, filed by en employee, whose cumulative trauma injury was not acknowledged by DuPont, as work-related.
>
> DuPont agreed to conduct a corporate-wide review of its injury, and illness records for the past five years, correct any deficiencies it finds, and compensate employees for any previous expenses incurred due to work-related injuries. (20)

Why? Why? Why?

In investigating what causes accidents, the three magic words—Why? Why? Why?—are invaluable. For every unsafe act, there is a reason. For every unsafe condition, there is a reason. These are called the basic, or the root, causes and should be sought, identified, and rectified. Only by rectifying the root causes will the repeat of a similar type of event be thwarted or stopped.

It should come as no surprise to accident investigators when they find that, in most accidents, people have committed unsafe acts. This is really a no-brainer. Thorough accident investigation will ask *why* the person committed the unsafe act. In an immediate cause analysis of unsafe conditions, the same question is asked: *Why* is this unsafe condition here? The risk of this basic cause analysis investigation technique is that it opens a can of worms and may reveal the truth. Not everybody wants to hear the truth. Wouldn't it be so much simpler to blame the accident on the unsafe act of the employee, take the necessary actions, and, after having gone through the motions, it's back to business as usual?

By delving into the basic causes of accidents, the management safety system is exposed, as are possible flaws in the leadership adequacy and style. What about the behaviors of managers? Who, after all, is responsible for providing a safe work area? Who, after all, is responsible for the actions of the employees?

Thorough accident investigation is precarious, and I warn people not to investigate accidents unless they want to get involved in an internal political uproar. The U.S. twenty-five-cent coin is unique. It has two sides. Accidents, too, have two sides. Thorough investigation will look at both sides and not only the path of least resistance or the unsafe act of the worker.

Managers manage human beings. Human beings will make mistakes. Because they make mistakes does not necessarily mean they are entirely to blame for their mistakes. If these mistakes are defined as unsafe acts, then management has an opportunity to train, retrain, guide, educate, and motivate employees to reduce the number and frequency of unsafe acts being committed. In numerous instances, employees commit unsafe acts merely by following supervision's instructions. That is another one-sided story. The supervisor says, "Get the production out," and the worker is virtually prevented from working safe. In some cases, he or she receives either direct or indirect instructions to take shortcuts, use improper equipment, and so on in an effort to maintain or boost the production numbers.

Loss Causation Models

In comparing the accident domino sequences of Heinrich and Bird, it is obvious that Bird has a point. If one looks at the science of management, it is quite clear that safety management is an integrated part of that business. Setting system standards and controls to reduce the risk is what prevents or reduces the probability of accidents. Bird also expanded the concept of safety by including all losses as a result of accidents. These included property damage, equipment damage, downtime, and business interruptions.

In *The Cause, Effect, and Control of Accidental Loss* (McKinnon 2005), I introduced the fifteen-domino model based on the work done by Heinrich and Bird. Research over twenty-six years indicated that the first event in the sequence of events, (an accident) depicted by the fall of dominoes, was "a failure to assess the risk." If the risk is not identified, it cannot be controlled. This, in turn, gives rise to the basic accident causes and then the unsafe acts and unsafe conditions. The three luck factors determine the course of events.

The unsafe act or condition may merely result in a near miss, which is luck factor 1. If there is a contact with a source of energy, luck factor 2 determines whether that contact is going to injure a person, damage property, or merely cause a business disruption. Should there be an injury as a result of the exchange of energy, luck factor 3 then determines the degree of severity of that injury.

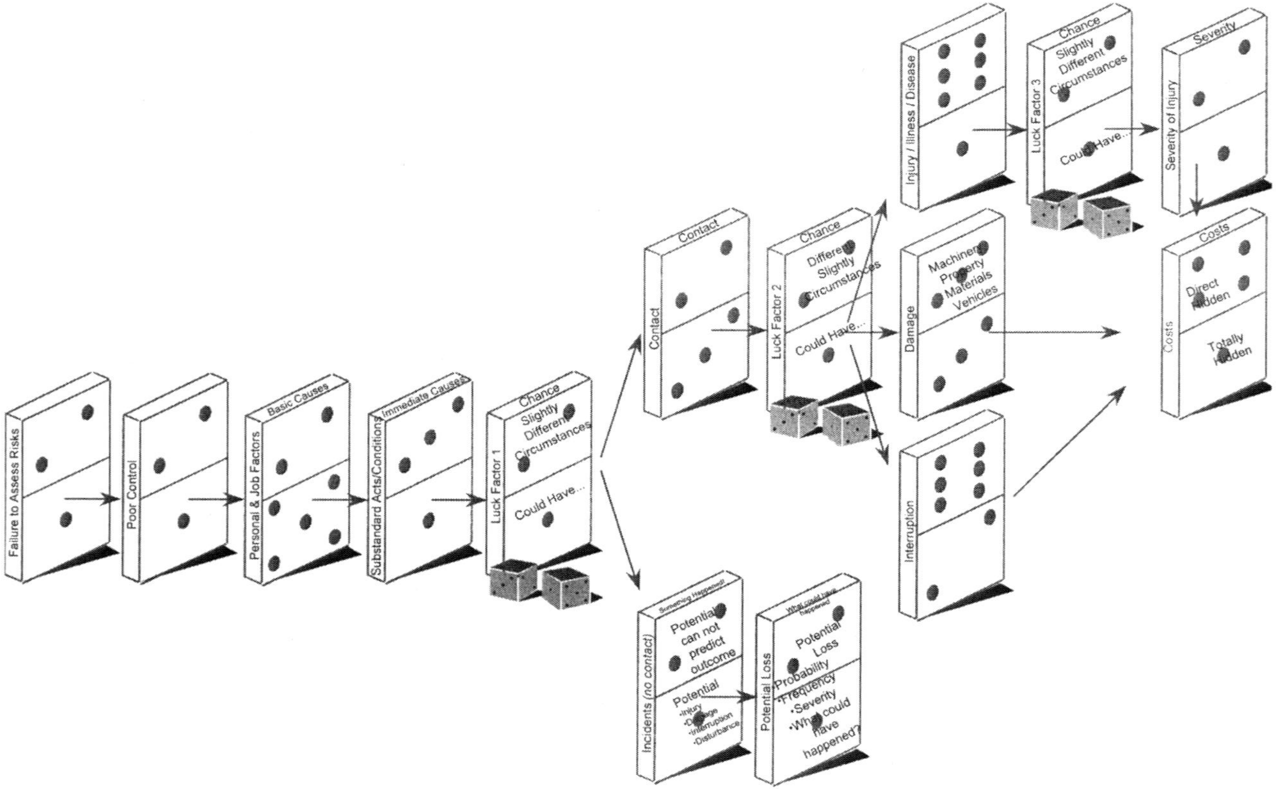

Model 4. The accident sequence showing the stages of an accident the path of which is determined by the three "luck factors." Copyright 2000 from *The Cause, Effect, and Control of Accidental Loss,* by Ron C. McKinnon, reproduced by permission of Routledge/Taylor & Francis Group, LLC.

Safety is entirely gauged, measured, and rewarded (or punished) on the degree of severity of the injury. Yet this resultant injury is a result of three luck factors, being three *bad* luck factors. If one is lucky, there is no contact, and only a near miss occurs. What determines the outcome of an exchange of energy: only luck factor 2?

Having researched numerous injuries, property damages, and thousands of near misses, I am convinced that the resultant injury is largely fortuitous. Heinrich stated this, Bird and Germain made the same statement, W. Tarrants also agreed, and so did safety professional Dan Petersen.

Heinrich's fourth axiom is that the *severity of an injury is largely fortuitous.* Yet the whole safety measurement scheme is based on these luck factors. We are really measuring luck. If someone works for a year without an injury, we automatically reward that person for working "safely." What about the hundreds of unsafe acts he or she committed during this period? What about the unsafe conditions they may have created? What about the negative safety influences they may have had on their fellow workers and colleagues? Yet we have rewarded the person for not being injured, so we are actually saying, "Well done, you were very lucky," but not necessarily *safe.* That reward reinforces the unsafe acts committed and the unsafe conditions they created. What do you think they're going to do the next year?

Conclusion

Instead of basing our safety actions on outdated and possibly incorrect premises and using these to abdicate safety responsibility and accountability, a more in-depth look at the real cause of safety problems, ignoring old-fashioned beliefs and myths, is needed.

6

Paradigm 5: Injuries Measure Safety Performance

FOR DECADES WE HAVE BEEN USING INJURY statistics to measure safety. The reason these statistics and figures are used is because the term *safety* has become synonymous with the term *injury*. These statistics, such as the lost-time injury frequency rate, disabling injury incidence rate, and injury severity rate, are really measures of injuries and not measures of safety per se.

Safety is the control measure implemented to reduce the probability of accidental losses occurring. Injuries are only one result of poor safety. We incorrectly continue to count the number of casualties and equate this to safety performance. That's not safety—that's a casualty tally.

Luck Factor

In analyzing any loss-producing event, there are at least three luck factors that determine the outcome of the event. Should there be an exchange of energy and a resultant injury from luck factors 1 and 2 (over which we have no control), luck factor 3 then determines the severity of the injury.

Even H. W. Heinrich (1959) included this statement in his ten axioms of safety when he said, "The severity of an injury is fortuitous" (28).

Leading safety authors have also proposed that for every serious injury, there have been numerous undesired events that could have also resulted in injury. So here we have a situation that ignores even the oldest safety philosophies. We still measure the number of reportable, disabling, or lost-time injuries and use this as a gauge, often the only gauge, for our safety performance. What we are really measuring is *luck* and not *control.*

Traditional

The using of fancy formulas and statistics depicting the injury severity and fatality rate is traditional safety. These are all downstream measures. What they measure is a failure in the management system. What serious injuries depict is less than 1 percent of the total picture. In referring to the Bird accident ratio, only 0.15 percent of undesired events result in serious injury. Yet the serious injury is what is used to measure and compare one company with another. It is what legislation uses to determine whether a physical inspection of the work site is warranted. It is the figure that international and national safety organizations use to qualify organizations for safety awards.

The injury rates are now quoted by organizations in their annual reports. This is admirable because the boards of directors are now reporting the state of safety to the board and the shareholders. However, when the chairman tells his shareholders that the injury rate is an "internationally accepted measure of safety performance," then you know you are in trouble. If the chairman is quoting the number of injuries, then the shareholders will also start to believe that the number of injuries experienced is a good measure of the company's safety performance.

Let me explain what a poor measure of safety the injury rate is. If underreporting in an organization is 50 percent, then only half the injury rate is depicted. The safety record could improve with the increasing dishonesty brought about through traditional safety competitions, incentive schemes, and disciplinary programs. These all artificially help lower (hide) the rate. What about the criteria for measuring these injuries? These differ greatly, and therefore the end results cannot be accurately compared. The players are playing the same game to different rules.

What Is a Lost-Time, Disabling, and Reportable, or Recordable, Injury?

If you are a safety professional employed by a large organization that is geared to measuring its safety performance by injury numbers and you are part of the number-juggling scene, then I suggest you stop reading immediately.

Why I say this is because an organization may be exposed when we consider the following questions honestly:

- What is the definition of a *lost-time* injury?
- What is the definition of a *disabling* injury or disease?
- What is *reportable* under OSHA (Occupational Safety and Health Administration) or local legislation?
- What is *reportable* under MSHA (Mine Safety and Health Administration)?
- What injury is reportable under your particular labor law?
- What *degree* of injury must be reported to your corporate office?

These are the burning questions. In any sport, the number of successes is measured against standard rules. In soccer, a goal is a goal. The rules are governed every minute of the way by a referee and two linesmen. All the players play the same game according to the same rules. There are matches where there have been exceptions, I agree. For the purpose of this discussion, let's have a look at the rule rather than the exception. Basketball is played to a set of rules and strictly monitored by umpires. Baseball is monitored. A home run has to qualify and meet a certain criterion before it is a home run.

Safety has hundreds of different rules that determine what level of injury counts and what does not count. Not only does safety have differing and conflicting definitions of significant injuries, but these definitions are also massaged, modified, manipulated, fudged, and cooked to suit the system.

We often compare apples with pears because of these conflicting guidelines and modification to injury guidelines. I guess safety measurement by using injuries as a yardstick is the only game in the world where you can write your own rules.

Many a safety practitioner makes his mark by cooking the books to make his organization look good. This makes him look good, and he receives a pat on the back from management. Meanwhile, back at the workplace, people are being injured. Juggling the figures happen so that the "safety record" can be preserved at all cost. Some safety practitioners are employed to keep the injury rate down. If they can't do that by discipline, persuasion, and appeal or by begging people to "have a safe day," they (some) then fudge the records.

Doctors

I have personally experienced the following. I visited the local hospital because of an ear infection developed while flying overseas. As I walked into the hospital, the doctor recognized me as being an employee of the safety department. He immediately took on a subservient role and approached me most apologetically about James Brown. (Not the person's real name.) I then questioned him. "Well, what about James Brown?" He said, "I'm terribly sorry, I've tried everything, but he will not be able to return to work for the next shift."

Being curious, I then questioned the doctor as to what had happened. James had apparently been rather severely injured in a fall at work and was having difficulty walking because of massive bruising of his hip and right leg. The doctor continued, in his apologetic manner, to explain to me that he just could not allow James to return to work. It then dawned on me that if James could have returned to work, it wouldn't have been counted as a "lost-time injury" but only as a minor injury.

Here I was in the hospital with a medical doctor *apologizing* to me for the fact that our organization had injured James so badly that he could not return to work the next shift. What am I trying to explain here? What I am saying is that some organizations, some safety departments, and some safety personnel almost *intimidate* doctors into releasing patients injured on the job back to work to preserve the safety record. Fudging the books is what many organizations do.

Defining an Injury

Some definitions require the injured person to return to work without losing a complete shift or more. Some definitions require the person to return back to "meaningful work." Some definitions, on the other hand, require an employee to return to "one of his regular established jobs." Some legal definitions require the man only to return back to the property. If the man does not return to the property after a complete shift, it is then deemed *reportable.* (This means the injury needs to be reported to the local labor authorities.)

Some legislation has an even finer dividing line by qualifying a reportable injury as any injury that required medical treatment by a doctor or the use of a medical instrument. A simple example is sutures. If a doctor sutures a wound caused by an accident, under certain legislation it is *reportable.* So guess what? The first-aid department uses butterfly clips instead to close the wound; no stitches are required, so it no longer becomes *reportable.* The record is intact.

The acid test for an injury that is significant is the following:

- What was the employee doing at the exact time of the contact and subsequent injury?
- Could he or she, after receiving medical treatment, return to the same task without losing a complete shift or more?
- If the answer is "yes," then the injury has not hampered the business process. If, however, he or she has to come back to any other task *other*

than the task for which he or she was being paid at the time of the accident, then the accident has altered the work process, and a change has had to be made. This is an injury, which has changed the business process, and is a genuine "lost time" or "disabling injury."

Self Truth Test

Apply these criteria to your own organization and then recalculate your (real) injury rate. Shocked? I'm sure you are.

Even the *Encyclopedia of Occupational Health and Safety*, produced by the International Labor Office (1983) states, "There is as yet, no international definition of employment injuries" (34). This means the traditional measurement of safety, the injury, can be manipulated from company to company.

One organization I know changed the definition of a lost-time injury from one day to three days off work to make their safety appear better than it was.

I'm Not a Lost-Time . . .

I remember having discussions with Bunny Matthysen, the originator of the National Occupational Safety Association's 5-Star Health and Safety Management System. Bunny is an inductee into the Veterans of Safety Hall of Fame. He was telling me that in the late 1960s, they were on a competition judging tour and, on walking into a mine workshop, encountered an employee lying on a stretcher on the floor. The employee was in a plaster cast from his heels to his neck as a result of a work injury. When the judging team asked the employee what he was doing, the employee said, "I'm counting out these nuts and bolts and putting them into the bin." This the employee said while lying flat on his back using his only movable limb, his right arm, to do the job. The employee hastened to add, "But I'm not a lost-time injury, I'm back at work!" This is an actual case study that exemplifies the fact that the statistics can, are, and will be fudged if they are used as the prime measure of safety.

What I am saying is that there is no standard, either countrywide or internationally, that clearly spells out the rules of the safety injury classification and counting game. The closest we come to that are standards Z16.1 and Z16.2 of the American National Standards Institute (ANSI). These are excellent rules and give examples and clear guidelines as to what is a disabling injury and what is not. The only pity is that nobody uses these to gauge their injury performance. The reason for this is that they are too strict, and injury rates will go through the roof.

Why Don't People Report Injuries?

I have posed this question to numerous trainees in my various classes over the past thirty years or so, and the most common reasons are given as follows:

- Fear of repercussions
- Fear of disciplinary measures being taken
- Possible loss of a safety bonus
- The person does not want to spoil the safety record
- The injury is recorded on their personal record
- Letting the team down
- Embarrassment
- Didn't think the injury was so serious
- Incurring the wrath of the supervisor
- Fear of getting "written up"

These are the most common reasons given in reply to my question. One particular organization actually posted the name of the injured person on the injury experience board at the entrance to the factory. This meant that when the number of "days without serious injury" dropped to zero, the "culprit's" name was also displayed. "*We lost our 900,000 injury-free (safe) man-hours because of Ron's injury!*" Imagine that being flashed on the board at the entrance to the factory.

Employees make certain that this does not happen to them and therefore do not report injuries. Fear of repercussions is perhaps the most prominent reason given for nonreporting of injuries. There are apparently severe repercussions for a serious injury. After an injury, the supervisor's performance record is tainted, the safety bonus is lost, and peers and colleagues become antagonistic toward the injured employee. After all, he ruined the safety record.

What is really ridiculous is that the injured employee could have been injured as a result of somebody else's doing. Another organization gave employees stickers for being injury free for each year. Employees were embarrassed not to receive their stickers and consequently did not report their injuries.

Fear of Reporting

In interviewing a group of twenty-seven managers, 90 percent admitted that there was fear to report an injury, as this was entered on the employees' personal records. A number of them stated that in the past, employees had been fired for having had a certain number of injuries. This leads to the suppression of the reporting of injuries and once again skews the statistics. Honesty in reporting injuries no longer exists because of these repercussions. Employees, unions, and union members see this ridiculous safety scenario, and consequently their disrespect for safety and safety management is fully warranted.

Black Marks

This nonreporting of injuries obviously skews the injury rates on which so much is based. If safety performance, recognition, and progress are based on injury and severity rates, the many reasons for nonreporting are producing inaccurate statistics. Therefore, how accurate are our national and international figures?

I attended a safety lecture during my first week of employment at a major organization. The lecture was aimed at employees and was about the consequence of being injured. Interesting, I thought, until the second slide came up explaining the consequences of being injured at this plant:

- Loss of the "gold card" (a cash award for safety)
- Promotion chances reduced
- Being branded a "black sheep"
- Not being chosen to work overtime in the future
- Chances of "acting" for the supervisor in his absence: slight

Needless to say, we had to do a lot of work to change the safety culture of this fault-finding approach to one of safety motivation and participation.

How Have Others Improved Their Safety?

Most organizations have improved their safety, or so they think, merely by driving the number of injuries underground. What I mean by this is that they have increased the awareness of what an injury means to them and therefore pressurized all and sundry not to be injured but to "work safe" and "think safe."

I have personally witnessed a number of cases where injured employees have gone to their private doctor, outside of work hours, and had the injury treated as a sports injury. Returning to work partially incapacitated as a result of a "home," "sport," or "recreational injury" puts the injured person on a pedestal, and his fellow workers treat him like a hero. After all, it wasn't a work injury, was it? Should he be injured at work, he would, conversely, be treated as a jerk.

Off-the-Job

In reviewing the 1997 injury and fatality statistics produced by the National Safety Council (1998), I wonder why there are 1.5 off-the-job injuries reported for every 1 injury on-the-job. During that same year, there were nearly 6 million off-the-job injuries and 4 million on-the-job injuries. Is the off-the-job injury rate higher because workers are taking their injuries home and paying for their own medical treatment with their own medical insurance rather than "spoiling the company's safety record"? Surely, American homes are not more hazardous than American workplaces!

Injury-Free Bonus

Some time ago, it came to my attention that numerous contractors pay their employees an extra dollar per hour worked for being "injury free." So let's take an average contractor on a major project working twelve hours a day, seven days a week. In one month, that employee would have worked 360 hours, and his bonus for being injury free would amount to nearly $400.

Take the scenario of a young twenty-three-year-old contractor on the last shift of the month working on a project. He is about to end the day's work and then shower and pick up his paycheck and head for the big city and the bright lights. An hour before the end of the shift, he slips, falls, and sprains his ankle. Do you think that young man will report the accident and subsequent injury to his supervisor and lose his bonus? Or would he rather hobble to the pay office, collect his paycheck, and enjoy spending the $400 safety bonus? You be the judge.

Paying people to be injury free drives injuries underground. Disciplining people for being injured also drives injuries underground. Removing certain privileges and openly ridiculing injured employees drives injuries underground. With all these injuries going unreported, management is lulled into a false sense of security and thinks the plant is safe. Meanwhile, under the surface, you have the walking wounded running the factory.

Insect Bites

We know that some events are beyond our foreseeable control, but, nevertheless, sometimes insects and animals do cause injury at work. Take, for example, the case where an employee was working at his desk and a spider crawled up his ankle and bit him. The employee immediately developed a rash, his ankle swelled up, and he had difficulty walking even to the first-aid station.

They immediately referred him to the hospital, where he was treated accordingly. Because of the reaction to the spider's venom, the employee was booked off work for two days. The local safety manager immediately stated, "We are not buying [accepting responsibility for] this injury."

According to the ANSI recording and reporting standards, insect bites, arising out of and during the course of normal employment, are work-related injuries. Should a shift or more be lost as a result of an insect bite, a lost-time (disabling) injury should be recorded. See how ludicrous this system is? Honest reporting and documenting will mean your organization will have a poor safety record. Now you can accept that or tweak the records to continue with the safety record.

Safety Is Legislated

Safety at workplaces is regulated by government-appointed agencies. Around the world, there are factory inspectors, regulators, other governmental and mine inspectors, and watchdogs ensuring that the employer looks after the health and safety of the employee. No other aspect of business, other than finance, employs so many watchdogs and passes so many rules as these safety watchdogs of industry and mines. Why? Because managers cannot manage safety. Line managers can get eight hours of production out of employees but cannot prevent accidents, many of which result in injury.

These legal entities also put final judgment on organizations based solely on the number of injuries reported to them as per their criteria. As discussed, these "criteria" differ from state to state, country to country, and nation to nation. Some legal entities carry out safety inspections of organizations that have high injury rates. So how does the firm avoid an unwanted visit and safety inspection by a legal inspector? It merely makes sure that the injuries are not within the reportable category. Pay employees to be injury free, and they will be. Come rain or shine, they will be injury free. The only injury they will report is one that is so bad that they are almost crippled and that they can no longer hide. I have interviewed employees who told me that this is what they will do. For an average of $150 a month, employees will hide injuries.

Experience has shown that once legal entities enter a plant that is experiencing a number of injuries, they immediately find numerous nonconformances to legal requirements. So an organization that does not comply and does not want to be inspected and wants the best of both worlds will make sure that the injuries do not surface. They use incentive schemes, competitions, bonuses, and everything else to drive injuries underground so that things appear to be going well. They try to hide the symptoms and never get to the root causes. The myth that safety performance and injuries are a direct correlation has led to this situation.

Citation Quotas

For the first time in twenty-seven years, OSHA legislation was passed in the House and Senate. Originally drawn up by Cass Ballenger (R-N.C.), the laws were passed in the House before being voted on in the Senate. One bill bars OSHA supervisors from setting penalty quotas for inspectors. So what was happening before?

Competitions

Let's pause here for a moment and discuss *traditional* safety competitions. Traditional safety competitions are competitions to see which individual, which department, or which division works the longest without a disabling or lost-time injury. As Mark Friend (1997) said, "A typical program rewards employees for hiding facts—by encouraging them to not report injuries. Conversely, sound safety incentive programs reward employees, who offer input that helps everybody operate more safely" (34). What Friend is saying here is that the more you pay for being injury free, the more employees will hide the facts.

I once interviewed employees of a large mine that had recently gone out of business and closed down. Another company had purchased one of the surface-based activities of the mine, as they were in the same business. One of the purchasing criteria was their previous safety record. According to the books, they had been almost totally injury free for the past fifteen years and had never experienced a fatality. Only after the purchase did the truth surface. The physical conditions were horrendous. The electrical reticulation system was dangerous, there was no semblance of housekeeping or demarcation, stacking and storage was poor, and the buildings were damaged, bent, and twisted almost everywhere as a result of forklift–truck collisions. In interviewing one of the employees, he told me that the mine had paid a "hefty" monthly bonus to employees if they were injury free. Need I explain further?

Safety Award

Another organization I consulted to awards their employees a magnificent safety belt buckle if they work five years injury free. One particular individual was a "safe" worker and had not been injured. One day, however, his fellow worker was clowning around and dropped a brick from a high scaffold, intending it to land next to him to give him a fright. Unfortunately, the employee started to move as the brick fell, the brick hit him, and he suffered a serious injury to his shoulder and collarbone. Needless to say, he did not get his belt buckle award, but guess who did? Yes, his colleague, who caused his injury, received an award even though he had committed an unsafe act that had resulted in serious injury. This type of safety award is unfair and is also based entirely on the luck factor.

The Injured Employee Is Penalized

In some instances, the injured employee is only a victim of the safety management process. The injury is the consequence of a failure of the management system. Maybe the injured employee is the honest employee. Injury-free employees may be hiding injuries and could be dishonest. The more the emphasis is placed on being injury free, the more ducks and dives will be made to ensure that employees appear to be injury free.

An injured employee should never be penalized for being injured. Even though he may have committed an unsafe act, it is only one of a number of causes of the accident. One should never assume that the injured employee is guilty until the final accident investigation report has been tabled. I have difficulty understanding why an injured employee is interviewed immediately after the accident while he is in the hospital or immediately on his return to the plant. Why is he drug tested? What about interviewing his supervisor? What about his peers and colleagues? Remember that the injured employee is honest. He is admitting that he was injured, but the injury is only a *symptom* of management failure. The basic cause of the failure should be identified rather than focusing on the employee who happens to have been injured.

Besides, when last was an injured employee's *supervisor* drug tested after an injury to one of his team? In a case of driving under the influence, do they test the passengers for blood alcohol level or the driver? Why should it be different at the workplace?

Management, Wake Up!

This is a personal appeal to all levels of management to wake up. My message is, *Stop using injury rates as the sole measure of safety.* They are the most unreliable statistics available. The information received is only part of the picture, and, where possible, people do manipulate the figures, especially if they are held responsible for safety. Rather, measure upstream safety measures, such as the systems, standards, and controls that are in place, to prevent accidental losses. Examples of these are the following:

- Has a risk assessment been conducted?
- What are the standards for housekeeping, machine guarding, lockout systems, demarcation, ergonomics, personal protective equipment, and so on?
- Is there a written safety policy?
- Do you, the manager, conduct a daily safety inspection of the work area? Do you chair the executive or midmanagement or other management safety committee?
- Do you let the safety person chair the meeting?

Safety is the thing that you do to manage the risks of the business. Wake up and start asking your safety personnel, "What ongoing activities are we doing on a regular basis to prevent risks causing losses?" Ask them that question and beware of the "safety flavors of the month." There are numerous little elements of safety on the market that people call a "program." These are not programs. These are merely elements of a program. Safety "bingo" and other gimmicks are no substitute for an integrated safety system.

A safety system should control at least seventy main areas of potential loss, and the company should have written standards that spell out the details of how to control those areas of potential loss.

Managers get what they want. Measuring safety effort is proactive. Counting the number of injuries is reactive.

As Charles Gilmore (1970) said at a National Safety Council Congress, "What is the sense of measuring, if a loss must occur before you act? That's reaction, not control" (29). As Dan Petersen (1998) said, "Quit looking at injury based measurements to assess system effectiveness" (40).

James Tye

James Tye (1996), the late director of the British Safety Council, stated that *injury frequency rates were measures of failure* rather than success. He further states that the underreporting of accidents in the United Kingdom runs at about 30 percent and estimates about 80 percent in some U.S. industries. In his conclusion, he says that injury frequency rates are "frankly irrelevant" (30). He also quotes Tom Peters as saying, "If you can't measure then you can't manage" (30). He recommends that industrialists should stop measuring injury frequency rates and start measuring the seventy-plus areas of a safety performance program. He concludes by saying, "Why do I quote these figures: Because it's hard to hide a dead body as opposed to a non-fatal accident" (30).

In a lighter vein, the following story certainly contradicts this statement.

A Student Relates . . .

During one of my safety presentations at a gold mine in South Africa, one of the students related a story to me. I can relate the story only as he told it to me. Whether it is true is another story. Life is cheap in Africa, and I wouldn't be at all amazed or shocked if this story happens to be true.

In the gold mining industry in South Africa, numerous awards and trophies are presented for *safety records.* One of the most prestigious awards is the McLellan Shield presented by the Safety Division of the Chamber of Mines. If my memory serves me correctly, this is the award presented for 1 million work shifts without a fatality. This is indeed an achievement in a deep-level gold mine in South Africa. Because of pressure bursts, seismic movements, and other events occurring in workplaces in excess of a mile underground, rock-falls, pressure bursts, and other devastating incidents are common. During 1992, some 800 underground miners were killed as a result of underground falls of rock, rock bursts, and so on.

The story goes that this particular gold mine had almost achieved its 1 million fatality-free shifts and was already starting to prepare for the festivities and celebrations that accompany this prestigious achievement. The mine employees were getting ready to attend one of the biggest parties the local community had ever seen. The contractors were busy erecting massive marquees and tarpaulins, and the caterers were already barbecuing sheep and oxen on the spit. Tables were set in the tents, and the red carpet was laid out. A stage had been erected with a sound system, and a table had been prepared for the dignitaries, who would fly in from the head office in downtown Johannesburg by helicopter to attend this prestigious function. Safety functions in South Africa are traditionally well oiled with plenty of beer and other booze. Other safety awards are also presented at this prestigious occasion, and these had been polished and were busy being displayed on the main table, on the stage, and in the large marquee. For the thousands of African workers, similar arrangements were being made, but in an adjoining area, as was the tradition in those days.

The million fatality-free work shifts would be achieved at noon on this particular day, and the function was due to start at 1:00 P.M. with the speeches, congratulations, and presentation of the trophies. Everybody was in high spirits, and there was a flurry to get the tent guide ropes adjusted and the carpets vacuumed and to set the tables with tablecloths and cutlery. Tables were reserved to seat dignitaries at the VIP tables, and all was in order.

Apparently at about 11:00 A.M., an hour before the millionth shift would have been registered, a contractor employee was busy knocking in a tent peg in the area of the main tent when he suffered a heart attack and died. The safety department was immediately notified, and the corpse was conveniently dragged under a piece of canvas and left there until after the million shifts had been achieved. Once the siren had blown, the bells had been rung, and the million fatality-free shifts had been registered, there was a flap and flurry as the accident suddenly *occurred*, and a fatality was recorded on the job, immediately *after* the million shifts had been registered.

This was the story. That is how it was told to me. An urban legend perhaps? When next you visit a South African gold mine, you make up your own mind whether it's true. The moral of the story is, "So who said it is hard to hide a dead body?"

Rewarded for Working Unsafe

Frequently, workers are even rewarded for working unsafe. A man receives an award for being injury free for thirteen years. Surely he must have committed some unsafe act or created one unsafe condition during that period. Yet he is rewarded. This reward is a strong message that says, "We reward you for creating unsafe conditions and working unsafe." The fact that he is being injury free is merely a factor of luck. He could even possibly have injured his fellow worker who now misses out on the award.

An employee once told me that he had received a certificate for having worked for three years without being injured. He accepted the certificate but tore it up even before taking it home. He admitted that he had been the cause of an accident in which his coworker had been injured. He also admitted that although he had been injury free, he had not worked safely for the three years.

Work Safely

On a point of clarity, working injury free does not mean that a person has worked safely. How can injury free be the determining criteria for safe work? The employee was lucky. Positive safety actions, contributions, and activities are what should be rewarded in safety. If safety is ever to become first, then the safety profession must redefine what "working safely" means. Maybe ninety years ago, to be injury free meant you were safe. This is no longer applicable today.

After Sixteen Years, He Was Injured!

I once attended a safety function where employees received longtime injury-free awards. Employees were called out to receive recognition in the form of a check ranging from $1,000 to $2,000, depending on the injury-free period.

One of the last employees who was recognized was an old-timer who had gone for fifteen years without being injured, but during his sixteenth year, he was injured when a concrete block fell from a scaffold, striking him on the hard hat and shoulder. The old-timer was a bit shaken up after the event but nevertheless continued his work shift after receiving minor first aid. This statement was proudly announced and followed by an apology that he could not receive his "sixteen years injury-free award" because he had been injured. After the function, I interviewed this old-timer, and, as I had thought, he had played no part in the accident whatsoever. He happened to be wearing all his personal protective equipment, working where he was authorized to work, and doing the job he had been instructed to do when the block fell on his head.

In this case, the injury-free awards had actually punished the old-timer because he had not been injury free. This was not his own choice, as somebody else had injured him. Again, it is repeated that awarding people for injury-free periods encourages dishonesty, is unfair, and does not mean to say they had worked safely (in accordance with standard acceptable safety practices) for that period of time.

7

Paradigm 6: Injury Free Is a Safety Record

ONE OF THE BIGGEST PARADIGMS of safety is that the fewer injuries you have, the better your safety record. This is not true. Injuries, as a result of accidents, are the end result of a chain of events that include three luck factors. Luck factor 1 determines primarily whether there is going to be an exchange of energy or merely a near miss. Luck factor 2 determines whether the exchange of energy is going to cause a business interruption, property damage, or personal injury. Once the exchange of energy causes an injury, the severity of the injury is then determined by luck factor 3. Therefore, what the safety record is based on is luck.

Dedication of Employees

The pressure put on the average employee to preserve the so-called safety record is tremendous. Believe it or not, some employees are very loyal and extremely dedicated to the organization. This is especially so when it comes to safety.

I once interviewed an employee who had an accident. He had tripped, fallen, and broken his ankle. The employee continued to work for six hours on his broken ankle. Eventually, he was so overcome by the pain that he collapsed and had to receive medical attention. His ankle was so swollen that the attendants had to cut off his boot with a pair of tin snips. Is this loyalty or an example of the fear of reporting an injury, letting the side down, and crashing the "safety record"?

Injury-free periods may be a result of dedicated employees such as this one. These may also be a result of disciplining an employee who has been injured. There may be other incentives that lead to underreporting. These boost the safety record, which is consequently false.

Peer Pressure

Peer pressure also rates high on the list as to the reason why injuries are not reported. Couple a cash bonus to injury-free periods and add peer pressure, and you have a certain formula for the safety record being preserved. Peer pressure also exists among managers who constantly compete injury rate against injury rate. This is an unfair way of competing, as different departments have different hazards, and therefore apples are not being compared with apples.

The Safety "Record" Board

Safety practitioners over the years have managed to convince organizations to erect massive safety boards on which the safety record is proudly displayed. "This organization has worked so many days without an accident." (This should read "without an injury.") The safety board is a 1930s advertising gimmick still used to promote safety today. Surely this is outdated. Should we not modernize and scrap it all together? Other boards read, "Date of last disabling injury" or "Department in the doghouse." Here the name of the department responsible for the last injury is displayed in the doghouse. Sometimes the injured person's name appears in the doghouse. What an incentive for nonreporting. I really wish these organizations would realize that safety is far more than being injury free. Safety is how good their managers can manage, not how long they've been lucky, unlucky, honest, or dishonest.

Severity of Injuries Depends on Luck

I must reemphasize here that the severity of the injury is fortuitous. H. W. Heinrich said it in his third axiom, W. E. Tarrants said it in his book, and other safety practitioners and pioneers have written about it in one publication or another.

Generally, they all agree: the severity of an injury depends on luck. In fact, H. W. Heinrich and E. R. Granniss (1959) state clearly that "there are certain types of accidents, of course, where the probability of serious injury may vary in accordance with circumstances" (28).

Here they are explaining that luck determines the circumstances. They continue by saying, "The severity, or cost, of an injury that results when an accident occurs is difficult to control. Therefore, attention should be directed to accidents as properly defined, rather than to the injuries that they cause" (28–29).

Tarrants (1980) states, "To the extent that it can be argued that all injury-producing events are the consequence of the same causal structure, and that the degree of injury is merely a matter of chance" (180). He also agrees that there are clear advantages in a measure that includes all events in the sample.

Tarrants further explains that a low injury frequency rate does not tell much about the safety controls, nor does it predict the potential for catastrophe that may exist. He continues his arguments and calls injury rates "too crude." He says that "the rate can fluctuate drastically on a fortuitous basis. Likewise, measures of severity can fluctuate widely by chance" (Tarrants 1980, 239).

The Health and Safety Executive's Accident Prevention Advisory Unit (2006) says that any simple measurements of performance in terms of injury rates are not a reliable guide to the safety performance of a business undertaking. They also state that there is no clear correlation between such measurement and the work conditions or in injury potential or the severity of injury.

Dan Petersen (1978) states, "Perhaps our biggest doubt comes when we ask ourselves whether a death count or death rate really reflects our national effort and performance in safety" (128). He agrees that we seem to believe that fatalities are caused largely by "luck," and thus we tend to raise our eyebrows at such a measure.

He continues his argument by saying, "But the fact remains. The supervisor of ten workers can do absolutely nothing for a year and attain a zero frequency rate with only a small bit of luck" (Petersen 1978, 66). He agrees that by rewarding such a supervisor, we are actually reinforcing nonperformance in safety.

The main question, one that will help put safety in the foreground, is, "Are you in control or just relying on luck?"

Examples

During the same month, two accidents occurred at an organization. A heavily built employee was climbing down a vertical furnace shaft when he lost his footing and fell sixty-two feet to the bottom. At the bottom of the shaft, there were numerous obstructions, piles of brick, and other paraphernalia. The employee landed on his right side, in the fetal position, fortunately missing all the objects. He landed flat on the floor of the shaft. He suffered only minor bruising to his right side and returned back to work the very next day.

In examining where he had fallen in relation to the piles of bricks, tools, and other items, it was almost a miracle that he had fallen perfectly in between these items and had not fallen on them. Had he struck one of these items during his fall, these would have inflicted serious injury.

Not long afterward and only about a half mile away from this furnace, another employee was three feet off the ground on a four-foot stepladder. He lost his footing and fell approximately three feet to the ground, landing on his back. Unfortunately, he landed on the curbing of the roadway, and this caused severe and permanent injury to his spinal column, paralyzing him from the waist down.

So here we have two undesired events. One *should have* resulted in fatality. The other *should have* resulted in a minor injury. Yet, because of the luck factors, the degree of injury was totally unpredictable and completely contrary to what one would have expected.

Record of Failure

A news broadcast on national television once quoted a large mining organization as saying, "The mine's safety record has been improving over the last few years." I then questioned on what grounds this statement had been made. On investigation, this particular mining group killed, on average, sixteen employees per year. When asked how they could justify the fact that they were apparently improving, their reply was, "Last year we only killed twelve employees." So this was their impression that their safety record was improving. Was their safety improving, or were they getting *luckier*?

Let's revisit the statement made by Charles Gilmore in a paper presented at the National Safety Council Congress. Perhaps the most significant statement he made during that presentation was, "What is the sense of measuring, if the loss must occur before you act? That's reaction not control."

Safety records are based on how many people get injured, the reduction in the number of people getting injured, or the hours worked without people being injured or killed. Are these measures of safety, or are they measures of failure? The minute the management system fails and causes an accident, injures people, causes property damage, or disrupts the flow of business, this is a *failure*. People call it a *record*.

Disabling injury incidence rates, lost-time injury frequency rates, lost-time injury severity rates, reportable rates, total injury rates and fatality rates, are all measures of *failure*. These are measures and tallies of the wounded and dead. Because so much emphasis is placed on them as *records*, many schemes are devised to preserve them by any method whatsoever. As James Tye (1996) said, "It's hard to hide a dead body as opposed to a non-fatal injury" (30).

Proactive Safety Records

Safety records should be the following:

- The tracking of the number of positive safety observations and rectifications that have been made
- The number of inspections that are carried out
- The number of trained safety and health representatives who are active
- The number of times that management chairs the safety committee
- The percentage allocation given for compliance to at least seventy elements of safety control
- The number of near misses reported

These are safety records, not the fact that an organization has been lucky and not injured anybody for a month or so. I agree that we will never divorce ourselves totally from measures of consequence, but it is time to start looking in other directions. We should look at the controls rather than the degree of consequence on which the entire safety management scheme is focused, adjudicated, measured, and rewarded.

Measuring Luck

Safety pioneer Dan Petersen once interviewed a supervisor of ten men who had reduced the injury rate from something like ten in the one year down to a low of four the next year. When he asked the supervisor what he had done or why his department's injury rate had improved, the supervisor apparently shrugged his shoulders and said, "Nothing. I did nothing differently. I guess I was lucky."

If you want to ascertain what the magic formula was that enabled one of your departments, divisions, or another organization to work a million manhours without a disabling injury, here's the trick. Ask the management of that particular division or company to list the management systems, standards, controls, and activities that they had practiced and operated during that particular period that had given them this injury-free status. If they cannot present you with a structured safety management system or risk reduction program, coupled with the necessary checks and balances, then they were only lucky.

One particular manager was asked what he had done to reduce his injury rate by some 50 percent during the last year. The manager shrugged and said, "Nothing. Luck I guess."

Rewarding Safe Behavior

Safety's big sacred cow is that most people believe that accidents and undesired events are caused by unsafe behavior. A simple deduction on their part is that safe behavior, which is defined as "injury-free behavior," should be rewarded.

Numerous tricks, gimmicks, and pat-on-the-back schemes thrive on rewarding "safe behavior." What these schemes really reward is the person's good fortune for not being injured or contracting an industrial disease.

Terminology

Even today, these schemes still refer to antiquated terminology, such as "lost-time accident." If they can't even get the terminology of lost-time *injury* correct, how can they then make a contribution to the reduction and elimination of risks? Many of these schemes focus on the person and not the problem. They play the man instead of the ball.

Some of these schemes boast about their achievements. "With these incentives, workers get a sweatshirt for 250,000 man-hours worked without a lost-time accident, a jacket for 500,000 man-hours, and a $500 savings certificate and wall clock for 700,000 man-hours."

Another organization gives gifts and individual recognition to associates achieving an "accident-free year."

Another scheme even boasts to give its employees a day's paid vacation for every six months worked without an accident. (They really mean a *disabling-injury.*)

I wonder what a thorough safety inspection and audit of safety systems, balances, and controls in place these organizations would reveal.

Bill Hoyle (1995) says that McDonald's restaurants have boasted on their signs that they have sold so many billions of hamburgers. Likewise, most large industrial plants display a prominently located *Mac Safety* scoreboard. The scoreboard tracks how many days have passed since a lost-workday injury.

Safety success is measured by how many million hours have been worked without an injury. Fires, explosions, and toxic releases are not included on the scoreboard, according to Hoyle. If you get to a safety target, such as 2 million hours, everyone gets prizes, such as a sharp safety award jacket. Gain-sharing and variable-pay awards are also based on having low injury statistics.

According to Hoyle, many workers know that the scoreboard injury numbers are often the result of creative bookkeeping and of assigning injured workers special light-duty work while they heal. (This is similar to the *Porta* Cabin story mentioned in chapter 17.) The numbers may be a fantasy, but the jackets are real, he says, so complaints are kept to a minimum. Workers are also keenly aware of the powerful peer pressure to not report injuries. Reporting an injury means you will be blamed by your peers for taking money and prizes out of the pockets of your coworkers.

Management recognizes that the problem behavior that they are most concerned with is to entice workers to think twice before reporting injuries, Hoyle (1995, 17) concludes.

Whistle-Blowers

William English (2006) sums up the alternatives: "Because whistle blowing is radical adversarial behavior, a decision to do so is a decision to resign" (55).

8

Paradigm 7: There's No Such Thing as Luck in Safety

A MANAGER ONCE TOLD ME that there was no such thing as luck in safety. That's the paradigm, the myth that we have all believed for so long. Unfortunately, the truth is that a lot of luck is involved in every accident and near miss. Proper risk assessment and mitigation actions are not luck. They are good safety management techniques. Identifying and rectifying basic causes of undesired events is good safety management. Where luck comes in safety is that the *end result* of an undesired event (be it an accident or incident) is largely the result of three luck factors.

Luck Factor 1

Once an unsafe act is committed or an unsafe condition exists, luck determines whether that situation is going to result in an exchange of energy or if the energy is going to miss the person, object, or body and result in a near miss or incident.

A very simple example is running a red traffic light. This is an unsafe act. Only luck factor 1 determines whether a person will get away with it (a near miss) or whether there will be contact with another vehicle. Every group I ask admits that the reason they got away with running a red light was because they were *lucky.* None can predict the outcome of an unsafe act or unsafe condition.

Luck Factor 2

If, because of bad luck, there is a contact and some form of collision with another vehicle or person, luck factor 2 determines whether the outcome will be a minor interruption of the journey, minor or serious vehicle damage, or injury to people. As bad luck will have it, the outcome could be all three.

Luck Factor 3

Should there be injury as a result of the exchange of energy, luck factor 3 determines the degree of injury. Admittedly, air bags, safety belts, and door strengtheners do reduce the degree of injury in vehicle accidents. These mechanisms are not safety mechanisms but are postcontact protection devices that reduce the severity of injuries by cushioning the passenger. Besides these items, the resultant injury is largely dependent on luck factor 3.

So, because safety is measured by the degree of injury and the degree of injury is determined by luck, safety measurements have become measurements of *luck*. Therefore, luck plays a major role in injury measurement.

Types of Near Misses

In modifying the Bird–Germain (1966) accident ratio slightly, the difference between a near miss with energy exchange and contact and a near miss with energy exchange and no contact has to be explained. For the purpose of expanding on the accident ratio, therefore, a *near miss* is when an undesired event takes place and there is an exchange of energy but no *contact.* That means that there has been an exchange of energy, something has happened, but that energy missed and there was no contact. This is largely because of the luck factor. An example is a chunk of concrete flying through the air and just missing a person by three inches. That is a *near miss.* There was energy exchange but no contact.

A near miss is also possible if there is an exchange of energy, there is a contact, but the energy level is below the threshold limit of the body or structure. In other words, no injury or damage is caused by the exchange of energy. These are referred to by Bird as "near accidents."

Here we have an undesired event, an *exchange of energy*, a *contact*, but no damage or injury. In other words, the exchange of energy is below the threshold limit and does no harm. An example is a chunk of concrete flying through the air, losing its speed and momentum, and falling on a person's foot. It has so little energy by this time that the contact is insufficient to harm the person's foot. This is normally also termed a near-miss event, although there was an exchange of energy.

The *accident*, as you will recall, is when the exchange of energy is *above* the threshold limit so there is *contact*; the energy is *greater* than the body, object, or item; and *damage* and/or *injury* occurs.

One Thousand Near Misses

The Bird–Germain ratio proposes that for each serious or major injury, there are 600 incidents with no visible injury or damage. He calls them *near accidents* or *close calls*. Based on my experience and research, it is proposed that there are in excess of 1,000 near misses as defined here.

These 1,000 near misses or close calls merely indicate how lucky an organization is. Over any sustained period of time, there are at least a thousand opportunities for injury and damage. Why don't these result in injury? Only because of the luck factor?

There are also plenty of unsafe situations for every serious injury. Unsafe situations are at-risk behavior and high-risk work conditions.

The 1,000 events, where there is potential for loss, do not result in a loss because of luck. Because of circumstances beyond our control, either the exchange of energy is insufficient to cause harm or damage or there is no contact. However, under slightly different circumstances, there could be major loss or what is commonly recognized as an accident.

For every one serious injury, there are *some* minor injuries, *more* property damage events, and *plenty* of near misses.

Traditional Measurement

Because of the luck factors involved in safety measurement, the measurement is therefore confusing, meaningless, and largely a measurement of luck. The future of the safety profession is to start measuring actual proactive safety activities that reduce risk.

The boards of directors of leading companies should set the trend by asking how many near misses have been reported and investigated rather than how many *hits* have been experienced. Proactive activities are upstream activities. Upstream measurements in safety are *before* the loss rather than *after* the loss. It is too late to measure once the loss has occurred. The degree of loss is largely fortuitous. Nearly all safety pioneers agree on this, yet we continue to measure the fortuity. This is another paradigm that must be changed.

Space Shuttle Columbia

According to the Columbia Accident Investigation Board (CAIB), the space shuttle *Columbia*'s demise was caused by a chunk of insulating foam that dislodged from the left bipod ramp of the external fuel tank and struck the leading edge of the left wing, causing a breach of the thermal protection system. This in turn allowed superheated air to enter the wing and melt the aluminum structure, causing the wing to fail until aerodynamic forces caused the breakup of the orbiter.

The CAIB also found that foam had been shed on more than 80 percent of the seventy-nine missions and that debris had caused damage on every space shuttle flight. There had been insulation shedding on most missions. The report also asks the question why the shuttle missions continued when they were aware of the foam being shed during ascents. The CAIB further stated that it would seem that the longer the shuttle program allowed debris to continue striking the orbiters, the more opportunity existed to detect the serious threat it posed. Were they perhaps relying on the luck factor?

9

Paradigm 8: Safety Is Common Sense

EVEN TODAY A NUMBER OF PEOPLE SAY, "Safety is really common sense you know."

Regretfully, I have news for all who are still locked into the paradigm that safety is merely "common sense," a word that has been used as a cop-out for a long time. Safety is no longer common sense. According to modern safety management practices, safety is no longer common sense but entails a clear understanding of risks at the workplace, how to identify them and reduce them, by applying sound management principles encompassed in a structured safety system.

After an accident, an employee is often accused of not having used common sense. Very few people can define common sense. As long as we constantly blame failures of the management system, which cause accidents, on the employees not using common sense, safety will never improve.

Examples of Safety Common Sense

Here are some true-to-life examples of how safety is not (so much) common sense:

- Apparently, a can of insect spray is not sufficient when tackling a wasp's nest. Many people like to do the job thoroughly, efficiently, and quickly but not so safely. Merely light a cigarette lighter a short distance in front of the spraying insecticide, and you have a most effective blowtorch. With wasps, you can never take chances.

- A contractor carrying out construction work on a welding site was discovered to be welding with no eye protection whatsoever. The electrode holder was a pair of pliers that had been connected to the main cable, and the eye protection that he was using was a black-and-white photograph negative.
- During an incident recall session, a candidate recalled how the washing machine kept on tripping the GFCI (earth leakage device/ground fault protector). He then disconnected the ground wire from the plug of the washing machine. The GFCI stayed in the "on" position and did not trip, and the washing machine worked. When I asked him how long ago he had disconnected the ground "safety" wire, he stated, "Oh, about six years ago, but the machine is still working!"
- "Oh! So you've never heard of acetylene bombs! We have fun with them in the workshop every week. You simply adjust your oxygen and acetylene-welding torch until you get a nice flame, extinguish it, and pump the flammable mixture into any plastic bag you can find. Once the plastic bag is full, seal it, and you then insert a piece of string or similar fuse, and you have an acetylene bomb. Sometimes they go off with such a loud bang that they shake the dust off the rafters. They also give the guys in the workshop one heck of a fright!" (As told to me by an artisan attending one of my classes.)
- A man who had recently attended a safety training course had to carry out work on his steeply pitched tiled roof. His training taught him that he needed some form of lifeline for fall protection while he was on the roof. He obtained a long length of rope, tied a nonslip bowline knot around his waist, and sought a suitable anchor. The only readily available hitching point was the fender of his wife's car standing in the driveway. He tied the end of the rope firmly to the fender and proceeded to climb up on the roof. Taking up the tension, he felt quite secure and started working. His wife, unaware that her car was being used as a lifeline anchor, jumped into the car and sped off for her weekly hair appointment. The safety-conscious roof repairer was dragged halfway to the hairdresser before his wife realized that something was wrong with the pulling power of her car.
- A father fitted new carpet mats to his car. A few days later, he discovered, to his astonishment, that the newly fitted carpet was jamming the accelerator pedal, causing the car to accelerate at full speed. He vowed to refit the carpet so that it did not foul the accelerator pedal. While driving, he merely stretched down and pulled the carpet back, freeing the accelerator. His son borrowed the car for the day having to drive some fifty miles for

a job interview. On the way, the accelerator pedal jammed. The son bent down to grab the carpet to free the accelerator and momentarily took his eyes off the road. When he looked up, he found that he had drifted off of the road, and he jerked the steering wheel to bring the car back onto the road. This caused the car to roll three times. Fortunately, he was not killed but suffered only broken ribs and a severe bruising.

- A child came home from school with lice in her hair. The mother promptly placed her in the bath and proceeded to wash her hair with a pet anti-tick and flea shampoo. The poison was absorbed into the child's scalp, and she died some two hours later.
- After heavy rains, a home owner was confronted with a tripping GFCI in the main distribution board. He opened the distribution board to check the circuits and discovered that the circuit breaker feeding the swimming pool tripped the GFCI. He switched all the switches off, gathered his tools, and proceeded to the swimming pool, where he noticed that the pump had become wet during the rain. While opening the electrical terminal cover of the pump, he suddenly found himself lying on his back some ten feet away from the pump housing. His wife had returned home and, on noticing that the lights were out, had switched on the electricity.

So much for the paradigm of safety simply being a matter of *common sense.*

Responsibilities of Management

Management has certain safety responsibilities:

- To provide a safe, clean, healthy, and pleasant work environment for employees
- To make employees thoroughly familiar with the safety standards and the hazards that may be present at the workplace
- To train employees and to make them aware of safety by means of safety videos, warning notices, safety campaigns, and so on
- To develop safety rules in conjunction with the employees and to enforce them
- To provide and maintain personal protective equipment and to ensure that it is worn
- To investigate all accidents and incidents with a view to determine the basic cause and to follow up by instituting remedial measures
- To establish a reporting system whereby employees have the opportunity to report substandard conditions and substandard acts

- To set a good example by wearing personal protective equipment when entering the work area and by following safety rules and procedures
- To provide ongoing support for the safety program by budgeting finance and by providing resources
- To conform to all legal safety requirements

Responsibilities of the Employee

An employee also has certain responsibilities toward his employer, himself, and his family:

- To obey all the rules and instructions issued in the interest of safety and health
- To use the personal protective equipment and other safety equipment issued and to ensure that it is not abused or interfered with
- To report all hazards, injuries, or damage-causing accidents promptly
- To do nothing that would jeopardize the safety, health, and well-being of himself or his fellow workers
- To conform to the requirements of local legislation

10

Paradigm 9: After an Accident, Blame the Injured Party (and Fix the Problem)

THE BELIEF IS THAT BY BLAMING an injured employee, the accident problem will be solved. This is paradigm 9. In some instances, the injured employee is always blamed for the accident, but the real problem is never identified and solved.

Presumption

I was shocked to discover that at one particular organization most of the supervisors immediately presumed that the injured party was guilty of causing the accident. Without exception, they immediately focused on what the person did wrong and agreed that the injured person had brought the pain and suffering on him- or herself.

Whatever happened to the fact that a suspect is innocent until proven guilty? In a court of law, this pertains, but safety is different. Safety seems to have become a prosecution of the innocent and, therefore, another reason why safety has such a poor reputation.

The injured person is often an innocent victim of an undesired event. In numerous instances, the injured person did nothing to contribute to his injuries. He may have been injured by the unsafe behavior of one of his colleagues. The injured party may have fallen into a trap set by somebody else. Without a thorough, unbiased accident investigation, one cannot point a finger at any individual, least of all the injured employee.

The basic premise of behavior modification programs is that the primary cause of accidents is worker error. This blame-the-victim concept provides little opportunity for effective accident prevention. Behavior modification does not focus on the fundamental safety problems that we face in the continuous process industry, says Bill Hoyle.

Employee Failed to . . .

I remember the case of an underground miner who was working in an area where there was no illumination whatsoever. The only source of light was from his cap lamp. The walking surface was full of mud and covered in six inches of water. One day while walking down this drift, he stepped into an eighteen-inch hole, twisted his ankle, and fell to the floor in agony. I was sent a copy of the accident investigation report. The portion titled "What Caused the Accident" read as follows: "Employee failed to look where he was walking and twisted his ankle."

I was rather amazed at this statement, as in the total absence of lighting, how could he see where he was walking anyway? Even with good lighting, he wouldn't have been able to detect the hole hidden under the surface of this muddy water.

In exasperation, I turned to the portion of the form calling for remedial measures. The portion read, "Steps taken to prevent a recurrence." The answer, "Coached employee to exercise more caution when walking down the drift." *Coaching* the man to take more caution? What about the hole? Nothing was done about the hole or the lighting, which were the real causes of the accident. The hole is possibly still there to this very day. The basic cause was never identified, but the finger was pointed at the employee.

Accident Prone

This myth is another cop-out for poor management. Should a person be injured more than once, he is normally termed *accident prone.* The bad news is that accident proneness has never been disproved up to now. The good news is that it has not been proved beyond a doubt, either.

Some parties believe there could be the propensity for certain individuals to be injured more often than others. On the other hand, some believe that this is pure chance. Scientific evidence can neither prove nor disprove this. So, this cop-out should not be used as an escape for an accident that could probably be traced back to a systems failure. The organization's safety culture with respect to safety decision making determines the incidence experience obtained, according to the Columbia Accident Investigation Board.

Perhaps the person who gets injured more often than his colleagues is the one closer to the point of action. Maybe he is the one who works harder than the rest of the team. He may even be the most honest team member. He may be the individual who follows the procedure and reports every single injury, no matter how minor.

The safety system puts people in an awkward position. Employees are told to report all injuries immediately, no matter how minor. When they do report these injuries, these are recorded against their name, and at the end of the month he or she receives an automatic label of *accident prone.* Some are labeled *frequent flyers.* Workers call this being "written up."

I once had a youngster stand up in my class and tell his story. He was close to tears when he related what had happened to him. He told us that he had followed rules and reported every minor injury to his supervisor. Because of the number of injuries recorded against the team, they were disallowed their monthly *injury-free* bonus payout check. This young man was ostracized by the group and was transferred to a different department. He was really upset about this and said that he realized he should never have reported the injuries. The following gives more insight into the concept (fallacy) of *accident proneness.*

Some Information on Accident Proneness

Introduction

This book, many other books, and the teachings of safety specialists clearly inform us that accidents are caused by multiple causes and are inclined to happen in rapid sequence. We nevertheless brand the unfortunate victim of these accidents as an individual who is *accident prone.*

Any compensation paid to an injured person is basically an apology and a poor substitute for not having made the effort to prevent the accident in the first place.

The first reaction after an injury-causing accident is normally finger-pointing. Either a guilty party is found, identified, duly punished, or reprimanded, or an excuse is found for the event. One of the most frequently used excuses is that the victim was accident prone.

Accident proneness has been one of the many excuses used in industry and mining for poor adherence to or nonexistence of adequate safety standards.

Definition

Accident proneness *is the ability of an individual to be involved in numerous injury-causing accidents at a rate higher than can be statistically attributed to chance alone.*

What Is Accident Prone?

We all seem to understand the labeling of a person as accident prone. Once branded, we believe he or she is a walking accident looking for a place to happen. He or she is a person likely to be injured on a number of occasions, in succession, as a result of accidents. Accident proneness is used very loosely to brand an individual who because of circumstances has been involved in a number of accidents over a period of time. It is also a very quickly and loosely used term for an individual if it appears that the individual has had more than his or her share of injuries.

For many, it was a dead end to safety solutions. Once the finger was pointed to this accident prone individual, no further safety solutions or remedies were possible. This person had been branded.

Accident Prone or Injury Prone?

When analyzing the difference between an accident, a near miss incident, and an injury, it becomes clear that in modern safety management terms, the term *accident prone* is incorrect. It does not refer to an individual who is prone to accidents at all but rather to an individual who is prone to injuries caused by accidents.

The definition of an *accident* is *an undesired event that causes bodily harm to people or damage to property or loss of process.* Accident prone does not quite fit into what an accident is. If one considers the definition of an *injury* as being *subject to an external source of energy greater than the body can withstand,* it becomes clear that we are really talking about injury prone.

Does this mean that the injury-prone person was involved in more injury-producing accidents than statistically acceptable and that the accident-prone person was involved in accidents that led only to property and equipment damage? Or was the individual involved in numerous near miss incidents where, had circumstances been slightly different, injury to the individual could have occurred?

In the past, the term *accident* meant that somebody had been *injured* as a result of the accident, and consequently the term *accident prone* should rather be redefined as *injury-causing accident prone.*

Research over a number of years has described accident proneness as a combination of human abilities that makes a person highly proficient in *attracting* injury-causing accidents regardless of the work environment in which he has been placed. In turning this statement around, it may very well be that a lack of human abilities, skills, or knowledge may contribute to this apparent attraction of accidents.

Basic Causes or Root Causes?

It is always easier to label the end result of an accident with a label such as injury prone, accident prone, or something similar to shift the emphasis away from the root causes of the accident.

In effective accident and incident investigation, emphasis is placed on both investigating and determining the immediate cause of all accidents and incidents and then striving (by a process of delving, questioning, and elimination) to find the basic causes of the event.

This system is well entrenched in a minority of organizations. If the immediate and basic causes of the accidents are not sought, the obvious way out would be to point a finger and label the victim of the injury, "the injured employee," accident prone.

A thorough and in-depth investigation of each incident or accident would possibly have indicated that it was caused by something basic, such as stress, drug addiction, alcohol dependability, or some other factor. This, in turn, could have led to the substandard act or the substandard condition (direct cause of the accident).

Numerous experts, such as Greenwood, Woods, Yule, Newbold, Farmer, and Chambers, did intensive research on the aspect of accident proneness. One asks the question, "Did they follow the accident sequence right back to the immediate and basic causes, and control weaknesses to trace the root cause of the injury-producing accident, or was the emphasis rather placed on the individual under question?"

Chambers and Farmer actually coined the term *accident proneness.* This may have led to a misconception among industrialists and could mean that accident-prone individuals can be identified and removed from the workplace and that this will solve the accident problems.

The questions that need to be asked in modern safety science are the following:

- Was he an accident-prone individual or an injury-prone individual?
- Was he always involved when near miss incidents took place?

- Would removing such an individual from the work environment bring about an immediate reduction in accidents that damage property and equipment and also incidents that under slightly different circumstances could have damaged property or have injured people?

Accident-prone labeling is basically blame shifting and does not treat either the immediate cause or the basic cause of any accident. Unless one has asked at least thirty questions to uncover the basic causes under the classification of "personal factors," one can never brand an individual. By merely shifting the emphasis from the root cause to the individual personality and personality traits, we are treating the symptoms and not the cause.

The injury aspect of the accident comes about when an individual is subject to contact with a substance or force greater than he can withstand. This results in the subsequent injury to the person's body or health. Would a person enjoy these unplanned or uncontrolled contacts? Is it psychologically or physically possible to attract these contacts? Or is the unfortunate, injured person not perhaps the victim of a sequence of events that constitute an accident?

Accident Sequence

The injury or occupational illness is the end factor in a sequence of events. This sequence of events starts with the following:

- Failure to assess the risk
- Lack of managerial control
- No safety and health program in place
- No standards set for safety and health conformance
- Inadequate compliance to these standards once set
- Poor managerial systems

These, in turn, cause *basic causes* in the form of personal and job factors that, unless controlled, lead to the direct causes of accidents. *Natural causes* also feature as basic causes but for the purpose of accident proneness are not discussed in this book.

Substandard (unsafe) acts and substandard (unsafe) conditions are a direct result of basic causes, caused by inadequate standards or inadequate compliance with the standards, supposedly laid down by management.

What causes the contact with a substance or energy source, in turn causing the injury or illness, are the *immediate*, or *direct causes.* These are defined as the substandard or "unsafe" actions of people and the substandard or "unsafe" conditions of the work environment.

The *contact* could be any one of the standard "accident types," such as the following:

- Fall on same level
- Fall from different level
- Overexertion
- Contact with temperature extremes, etc.

These contacts, in turn, lead to the physical *injury*, which could be the following:

- Rupture
- Puncture
- Burn
- Asphyxiation
- Loss of a limb or part thereof, etc.

Once one takes the accident sequence into account, is there really a group of individuals that can be labeled, with all conviction, "prone to injury as a result of accidents on more occasions than is statistically expected by chance alone?"

Individual Differences

It is clear that there are differences between individuals, both physically and psychologically. This must be taken into account when employing people in a work environment and also from an occupational safety and health viewpoint.

These individual differences show that individuals have different perceptual reaction times, motor reaction times, and also different rates of learning, assimilation, and adjustment to work environments.

Sensory discrimination and muscular reaction are more influenced by hereditary factors than by environmental factors. Numerous studies have concluded that training could have little or no effect on either hereditary or environmental factors. In the early days, it was the order of the day to blame accidents on inborn factors, and this possibly gave rise to the term *accident proneness.*

No evidence has yet been submitted that the environmental conditions, experienced by so-called accident-prone individuals, were similar. Nor has any one individual been statistically proven, beyond a reasonable doubt, to be injury prone. The accident-prone-labeled person could perhaps be a victim of the work circumstances and the individual's own physical, mental, and psychological shortcomings.

Scapegoat

One normally believes what one wants to believe. Has the term *accident prone* not become the great safety cop-out or the great scapegoat for inadequate managerial control in not setting standards and managing toward compliance with standards? Is accident proneness another paradigm?

The word *accident* itself has become a scapegoat in many countries. By merely overshadowing the accident with the humble apology, "Oh, I am sorry, it was an accident," or shaking one's head in sympathy and saying, "What a tragic accident," all is apparently forgiven.

It has been indicated on numerous occasions that 98 percent of accidents are *caused* by immediate causes, which, in turn, are *created* by basic causes. These can be *identified*, *rectified*, and eventually *eliminated*, thus reducing the number of unplanned events that occur within a work environment.

The latest trend in the safety profession is to call everything an *incident*. This adds more confusion to the profession. An *accident* is still an *accident* regardless of how misconstrued the word has become.

Error-Prone Environment

If one indicated that accident-prone people came out of an error-prone environment, it would sound far more credible. According to some experts, a large percentage of all injury-causing accidents occur as a result of the substandard, unsafe, or error-prone work environment. Some 96 percent of all accidents never result in injury. Some are caused by acts of nature, and some cause property damage. It would be interesting to ask how many of these accidents that caused neither personal injury, property damage, nor production disruption involved the so-called accident-prone individual.

Even if a safety professional, with many years of experience, were placed in an error-prone environment, would his chances of being injured be statistically higher than if he were in a safety-prone environment?

Injury Reporting

The frequency of reporting near miss incidents, damage-producing accidents, and minor injury-producing accidents is poor in most countries.

Perhaps the accident-prone-labeled individual is that honest individual who reports all injuries regardless of how minor they may be to his supervisor, who, in turn, at the end of the day, produces a list of injuries caused by accidents, and the honest individual is labeled accident prone.

Definitions

In defining accident proneness, one must ask what standards are used to determine the definition.

What is the interpretation of an undesired event, a physical injury, an occupational disease, a minor injury, or a property damage incident?

Standards throughout the world differ. The term *accident* and its definition also differ. People refer to accident when they mean injury. We must talk the same language before we can term employees *accident prone.*

What about the Work Environment?

Work environments differ from industry to industry, from mine to mine, and from culture to culture, not to mention from country to country. What sort of work environment are we placing these people in? Could the environment be conducive to producing numerous incidents, accidents, and injury-producing events? Perhaps the experts should move the focus from the individual to the environment, which, as mentioned, could be directly responsible for the majority of all injury-producing accidents. As Bill Hoyle said, "Fix the workplace, not the worker!"

The reporting techniques, the depth and honesty of reporting, and the definition of what an injury is could have a marked effect on the criteria for labeling people as accident prone.

Semiskilled Workers

If an employee is consistently given unsafe procedure instructions and carries them out to the letter of the word, he or she will definitely be involved in accidents and more than likely injured. Lack of knowledge, lack of industrial experience, and the adherence to instructions by certain semiskilled employees could place them in a situation where they could be injured more frequently than is statistically expected.

Illiteracy and unemployment lead to a virtual unending supply of cheap, unsophisticated labor. This could result in certain individuals being involved in accidents on repeated occasions. Is this the individual's fault? Is this perhaps not the old military adage of "cannon fodder?" Were the hundred and thousands of troops who marched toward the enemy lines without any form of protection in a similar scenario? They were mowed down by the thousands and listed as victims of the battle. They were victims of poor military command as they walked into the enemy's fire, literally walking to their own deaths.

If one compares these scenarios, it is obvious that the term *accident proneness* has become a cover-up, a myth, and an excuse.

The Injury Is Punishment Enough

Another ridiculous statement in safety is that the injured person receives punishment for his actions. What this means is that if a person is injured and experiences pain and suffering, this is payment for his unsafe actions. I find this rather poor, childish, and extremely antiquated thinking. No one makes up his or her mind to work unsafely and be injured. After interviewing thousands of employees, middle management, supervisors, safety coordinators, and safety representatives, I have yet to meet an individual who wanted to be injured while carrying out a normal day's work. There may be exceptions to this statement, but the general rule is that people don't plan to work unsafely. The system allows or forces them to work unsafely.

Interviews with injured employees have highlighted the fact that the pain and suffering that they endured—the disruption to family, church, social, school, and home life—has been devastating. No amount of money could ever repay that breech of their lifestyle.

Would those individuals not become more safety aware or safety conscious? Possibly yes, but are we saying that an individual willfully puts him- or herself into the firing line to accept the physical and mental punishment? Is this done with the knowledge and understanding of the disruption of normal life that an injury-producing accident can cause? I don't think so.

Barriers to Safety

In certain countries, the following could pose a barrier between safety and the individual:

- Language barriers
- Cultural barriers
- Political barriers
- Religious barriers
- Previous work experience barriers
- Personal beliefs
- Fatalism
- Attitudes

With these barriers existing, the maintenance of safety is extremely difficult.

If an employee has a fatalistic approach, the safety effort centered on that employee may very well fail. Some of the previously mentioned factors may very well contribute to this employee being involved in more than one accident and being injured more than once while carrying out his normal work.

Carelessness

Carelessness has also been used on numerous occasions as the one and only cause of all accidents, whether they are injury producing, damage producing, or cause some other form of interruption.

Carelessness will never be found as either an immediate cause or a basic cause in the accident sequence. It is seen as a blanket to be thrown over the accident causation factors to merely cover the truths of the direct and indirect cause. This blanket obscures the truth and blots out the details that may have highlighted an unsafe physical or mechanical condition or substandard practice caused by a basic cause.

Carelessness does not rate as a contribution to accident proneness, but the two facts are often linked together.

Lack of Concentration

When looking beyond the simplistic term of *accident proneness*, one should look at other aspects, such as the mental concentration of people.

The concentration factor is highly important. Accidents are caused when people do not keep their minds on their work, and a lack of concentration occurs. The criticality of the job determines the degree of concentration needed. Concentration is the ability to take one's mind off the main idea and to think of other related ideas, one at a time, and then to return to the main idea without deviating substantially. Should the mind not be able to return to the main attraction, mental contact could be lost with the subject, and an unsafe act could result.

Lack of concentration could be because the person has too many interests and the mind is not operating at peak efficiency, as the activities are too diverse. These reduced powers of concentration may also be a result of a boring job and the fact that the work content does not stimulate the employee. It is necessary, therefore, to attempt to develop and maintain job interest for all employees and to train and ensure that the employees are interested in their particular job and understand its importance.

Stress, anxiety, worry, and fear are also influences that distract the worker's mind from the task at hand and lead to a breakdown in concentration. This creates mental turmoil, disconcertedness, and decreased efficiency that increases the probability of accidents occurring. These can be clearly defined as basic causes of accidents and certainly not an excuse to label a person accident prone or accident repeater, which is synonymous with *accident prone.*

Research Studies

Studies conducted by Arbous, Kerrich, and Schulzinger indicate that accident proneness is not a major factor in safety problems. Their report indicated that people who consistently experience injuries annually over a three-year period accounted for only some 5 percent of all accidents. The study also showed that some 74 percent of accidents experienced were due to relative and infrequent solitary experience of a large group of persons. A definite correlation did show up in the study, however, in that it appeared that the susceptibility to having accidents was reduced with increasing age of the groups studied.

Physical Factors

Rather than pursuing the myth of accident proneness, it would be far more beneficial to investigate the other unidentified physical deficiencies of the individual in employment. Look at other factors, such as vision, reaction time, hearing, age, work experience, emotional stability, susceptibility to fatigue, and environmental factors such as lighting, ventilation, ambient noise level, and general state of the work environment. Rectify any deficiencies found rather than label people as the cause of the accident that caused the injury.

To Err Is Human

Bill Hoyle (1995) quotes the Chemical Manufacturers Association as saying that "Enlightened managers realize that . . . most mistakes are committed by skilled, careful, productive, well-meaning employees . . . human error is a natural and inevitable result of human variability in our interactions with a system" (10).

Blame the Deceased

It is amazing how a fatal injury accident investigation tends to put blame on the deceased person. Is this because the victim cannot answer back or give testimony as to what happened during the accident? Is it because of a fear of litigation or repercussions after the accident? I believe the legal system is what clouds effective accident investigation and makes a mockery of safety.

Management, unions, employees, and boards of directors are more worried about legal litigation and being sued than about what injured the person. This is why there is a constant attempt to blame the employees who are injured. The deceased person is an ideal opportunity to download all blame. After all, he's dead; he must have done something wrong. What a simplistic approach to justify poor leadership.

Fatal Accident Investigation

Several years ago, I was assigned to investigate a fatal accident at a remote work site. The director of safety for the group joined us on our five-hour flight to the site of the accident. This was the first time I had met the director, who had recently joined the organization.

On arriving at the scene of the accident, he casually walked around inspecting the site without as much as a pen or piece of paper in his hand. He did not even have a camera or video recorder to take pictures of the accident scene. I took some thirty pictures, numerous measurements, dimensions, and about four pages of notes.

I remained on the site for the following six days and revisited the scene of the accident on six different occasions until I found the evidence I was looking for. The evidence I eventually found was a clear indication that the unsafe practice that had led to the death of the employee was not an isolated case. The unsafe practice had been carried out by numerous workers on the site that day and had become condoned practice.

I was called into the general manager's office halfway through the investigation and questioned as to whether there must *always* be disciplinary action after a fatal accident. I told the general manager that applying disciplinary action would be an emotional move rather than an action to prevent a recurrence of the event. He apparently was being pressurized by the board of directors to take some disciplinary action against someone to prove I don't know what.

After two weeks, I had completed a thirty-page accident investigation report that included copies of witness interviews, photographs, sketches, and drawings. A comprehensive immediate-cause analysis was included, as was a thorough basic-cause analysis. There were some ten immediate causes in the form of unsafe acts and unsafe conditions and about twenty basic causes that had led to the immediate causes. I coupled the investigation to the accident causation model. This model also clearly indicated three luck factors that had played a role in the outcome of this accident.

A day or two after submitting this report to the safety director, he requested me to "review" my investigation. I didn't quite understand what he meant by "review," but I reread the report, made some slight modifications, and ensured that the immediate and basic causes had been clearly separated and identified. A day after resubmitting the review, he called me and was obviously disappointed that I had not made major changes to my investigation. I explained to him that the causes of the accident had been clearly identified by research, interview, inspection, and retest and that I didn't know how to further "review" my findings.

This draft report was submitted to the management of the division with a message that read as follows: "The basic causes of this accident should be ignored, as they are only assumptions and not based on fact." This staggered me, and I realized that he did not want a truthful accident investigation report. What he wanted was a report blaming the deceased employee for his "unsafe actions." This statement would have exonerated all the management, the victim's supervision, as well as his colleagues.

I did not change my report, and when the safety director's final report was issued, it had been rewritten in a different format but contained 90 percent of what my original report contained. I made the mistake of not putting blame on the employee, as is typical in the safety industry. It is typical and a cop-out because the dead person cannot defend his rights.

Strange but true, about three months after the accident investigation, I was transferred to another division and placed in a rather insignificant position in the two-man training department. I referred to the new office as "the office without any windows."

I often wondered what led to my demise. I can conclude only that doing a truthful, accurate accident investigation and not blaming the employee can get you into hot water.

About a year later, the entire safety department was signed up to attend a national safety and health congress at a distant location. All received booking confirmations and airplane tickets except for me. No fewer than twenty-five people in the safety department were signed up for the conference. I was the only one excluded. This safety director was obviously not charmed or pleased with the fatal accident report.

System Failure

Every accident that produces a loss is a result of a failure of the system. If a person is injured, then the system has failed the person. Numerous people say, "But the man took it on himself to work unsafe." This is no longer valid.

The system, culture, values, and norms of an organization determine how a person thinks, behaves, and works. We should remember that managers get what they want. If you put a man in a pigsty, he will think like a pig. His quality will be poor, his attitude will be less than perfect, and risk of accidental injury will be high.

Conclusion

In conclusion, the injured employee is often incorrectly and unfairly blamed for the accident. Accident proneness, as a blame mechanism, has yet to be proved as one of the basic causes in the sequence of events that lead to an accident, a contact, and consequent physical injury.

Accident proneness is still a scapegoat for lack of adequate safety program standards, noncompliance with standards, or noncommitment by the management team. It is often used to transfer blame for accidents.

No individual has officially been labeled as an accident-prone individual. No cause of death has ever been ascribed to accident proneness. Until the concept is proved or disproved by the relevant experts and associations, it must be accepted that there *are* both direct and indirect causes of all accidents. These causes can be identified and modified to reduce the possibility of the contact that causes injury.

Hoping to rectify the accident problem by blaming the injured worker and focusing on his behavior is a paradigm. Using accident proneness and other cop-out paradigms is no excuse for poor safety performance.

11

Paradigm 10: If It's Unsafe, You Can Shut It Down

ANOTHER ONE OF THE "WARM AND FUZZIES" of safety is the myth that if a situation is unsafe, anybody can shut it down until made safe and there will be no repercussions. This is also part of some safety legislation but belongs in safety folklore.

Fallacy

The statement "If it's unsafe anyone can shut it down" is really a safety fallacy. In the thirty years that I have been involved in safety management, I have met only one person who had the courage of his convictions to shut down a process for safety reasons. The average employee will not risk his job by shutting a process down—safe or unsafe.

I was once in the smelting area of a large iron and steel smelting and casting organization when one of the supervisors whom I had met previously approached me nervously. He told me that he had taken a decision to shut down one of the heat converters because of a safety situation. Bearing in mind that this was a Friday afternoon and he was due to end his shift within an hour or so, I asked him what the situation was. He explained that this particular converter had been overcharged and was "spitting." This meant that, because of the overcharging, it was spewing sporadic spurts of molten metal out of the opening. He said to me that it was not worth the risk to his work team and that he had taken the decision to shut it down. He then said to me that he was extremely worried because they may not meet the weekend production quota because of the loss caused by the shutdown. This was the first individual I have ever met who had actually shut down a major process because of possibility of injury.

It so happened that I followed up on this unusual situation. On the Monday morning, on walking into his office, I found him grinning from ear to ear. When I inquired, he told me that despite the loss of one converter, they had made record tonnage during the weekend. I was flabbergasted and congratulated him on his good management decision. As I have said, this is the first time I've ever seen anybody shut down a process because of safety, and in this case, it had paid off. What would have happened if they had fallen short of their weekend production quota?

A recent poll conducted by a safety organization revealed that 80 percent of the people polled would not interrupt their work in progress to rectify an unsafe work condition.

Why don't people stop unsafe situations? The truth is that they are scared of losing their jobs. "Unsafe" is very difficult to define because of the three luck factors that determine the end result of a situation. Nobody can predict for sure that somebody is going to get injured or that property will be damaged. Shutting it down is really taking a gamble with no guarantees. Asking employees to shut down unsafe operations or situations where they are put in danger would mean literally shutting down half of industry.

Forklift Truck

One successful organization had appointed and trained safety and health representatives for every division. The main objective of these safety representatives was to be a safety resource to the department.

A recently graduated safety representative encountered a forklift truck with a crack in the weld of an upright of the overhead "headache" rack. I asked why it was called a headache rack and was told that if anything should fall from above, the driver of the forklift truck would not get a headache.

The forklift truck was operating on a steel floor in a building and was transporting six-foot lengths of metal approximately forty feet to a storage rack. On encountering this hazard, the safety representative immediately called the supervisor and instructed him to "shut the forklift truck down." When asked why, the representative pointed out the crack in the weld of the one upright. His words were, "The rack is unsafe; shut the forklift down." The supervisor turned to the safety representative and said that he understood that it was a safety issue, but could the forklift truck not finish the loading job, which will be complete within half an hour or so? The safety representative became agitated and said, "This is a safety issue; shut it down!" By raising his voice, the supervisor immediately began to react defensively and once again said, "It's a safety issue, sure," he replied, "but can't we finish the job first, and then I'll get it fixed?" The safety representative then said angrily, "You see, you guys always put production before safety; safety is never first and never will be first!" He then stormed off in a huff.

What was wrong with this scenario was that the safety representative was trying to make a point on an emotional level without consulting the first principle of safety: identify and analyze the risk. If he had done that, he would have realized that by looking at the *probability* of an accident occurring, the possible *severity* of the consequences, and the *frequency* of it recurring, the crack in the weld was an extremely low risk.

Sure, the supervisor could have finished the job with no risk to the operator, the forklift truck, or people in the immediate environment. Giving people authority to shut situations down should be accompanied by the correct training in risk assessment and tact as well as good communication methods.

Figure 2. It took a consultant, fifteen senior employees, and two safety officers three months to shut down this unguarded conveyor.

Guts

It takes a lot of guts to shut down an operation because of hazards. These hazards could be either the unsafe behaviors of people or the unsafe physical or mechanical environment. Asking the average employee to do so is really pushing the envelope and also putting him in an awkward situation. Giving every employee that authority means that you can hold him or her accountable if the job was not shut down. So, if an employee does not use this blanket responsibility and authority to shut a job down because of the risk and something happens, we can always hold him accountable and responsible.

This is an absolute cop-out and puts all employees into a lose-lose situation. If they do shut the job down, then their supervisors will immediately black mark them for being a troublemaker or a "safety geek." If they don't shut the job down and something happens, then they will also be in trouble.

Supervisors and managers should be the ones to take the responsibility of shutting jobs down, if they are unsafe. They should be there, where the point of action is, and they should decide whether the situation is safe. Giving this authority to each and every employee is putting him or her in a disadvantaged position.

12

Paradigm 11: Accidents Are Investigated Properly

ONE OF THE BIGGEST FAILINGS in the safety profession is accident investigation. Generally, accident investigations are carried out poorly and are mostly fault-finding exercises. This sends a negative message to the workforce and constantly reinforces the fallacy that management is almost always perfect and never does anything wrong; rather, it is always the injured employee who is at fault.

Guilty Party

Most accident investigations seek a guilty party. Teachings in safety over the past seventy years have emphasized that accident investigation is a fact-finding mission and not a fault-finding mission. Nobody follows these guidelines, and everybody tries to put blame on others for the accident.

This is one of my favorites and in this context is worth mentioning again. I once investigated an accident in which an underground miner had twisted his ankle badly and had to be evacuated from the underground workings on a stretcher. The particular working area had no lighting whatsoever, and employees were reliant only on their miner's cap lamp. The walkway was covered with three to four inches of muddy water, and while walking down the drift, the employee had stepped into an eighteen-inch-deep hole hidden under the water.

When the investigation was completed, the main cause of the accident was listed as "Employee failed to look where he was walking." In that situation, it was impossible for anybody to look where he was walking because of inadequate lighting. Furthermore, who could have identified the trip hazard under the muddy water? The remedial measures taken, as listed in the same accident report, were "Put the employee through a safety coaching and counseling session." I was appalled at this particular investigation but was later to discover that all investigations were similar. They were fault-finding missions and not fact-finding missions.

Safety Cop-Outs

During a particular assignment in which I had to improve the accident investigation system, I picked up some of the major cop-outs or blame-fixing statements used during accident investigations.

Carelessness

One of the most prominently used cop-outs was *carelessness.* In reviewing the accident investigation report, carelessness was given as the cause of the accident. There were no unsafe conditions listed, no basic causes, and no remedial measures. The report merely stated that the man was careless, and that was what caused the accident. To justify the pain and suffering of an employee, accidents must be investigated thoroughly. Pointing a finger and telling an employee he was careless is a cop-out and will not fix the problem at all.

Expect the Unexpected

My favorite cop-out for an accident is *employee failed to expect the unexpected.* I have personally seen this on hundreds of accident investigation reports. When I questioned the investigators as to what they meant, they told me in no uncertain terms that the employee should have been aware of his surroundings and should have *expected the unexpected.*

Inattentive

In other instances, the employee had been accused of being *inattentive,* and this was listed as the cause of the accident. Investigating accidents and using such generic statements as this, once again, sends a very clear message to employees that safety is not a priority. If safety were a priority, the immediate causes of the accident would have been sought, the basic causes found, and positive steps taken to prevent a recurrence.

Holydaze

One particular group of accident investigations listed the cause of the accident as *holydaze.* This was a new one to me, and I had to get an explanation. The one supervisor willingly provided that. His explanation was something like this: "Oh, holydaze, yes, that's what caused the accident. The employee did not have his mind on the job. He was busy thinking about the Thanksgiving Day holiday [which was coming up]. He had his mind on basting the roast turkey, roasting the potatoes, entertaining his family and friends, and enjoying the holiday. He did not have his mind on the job but rather on the holiday." I gathered that what he was saying was that the employee was in a "daze" because of the holiday. I then figured out that this was the meaning of "holydaze" and was, according to them, a major cause of the accident.

Lack of Present-Moment Thinking

One of the most frequently used excuses that I found on another batch of accident investigation forms was *lack of present moment thinking.* What does this mean? What the investigator was trying to say was that the person was mentally "not there" at the time of the accident—that, like "holydaze," he suffered from a temporary blackout or loss of consciousness. Does this mean that the person may have dozed off during work hours or that he went into some kind of coma in the middle of the shift? If this is the case, I would question the leadership ability of his supervisor.

The level of safety at the average workplace is okay until there is an accident that causes an injury. Then the investigator blames the employee, but meanwhile the immediate and the root causes of the accident are never identified or rectified. Accident investigation is one of the more than seventy critical control elements. Poor investigations send a very clear message that safety systems are not in place if the accident is not used as an opportunity to fix the real problem. The problem may, in turn, cause other accidents. What about the other control elements? Are they in place, are they according to standards, or are they nonexistent? Admittedly, it is easier to blame the employees for "lack of safety."

Employee Failed to . . .

Another excuse that appears frequently as the cause of an accident is *employee failed to size up the job.* This was an excuse dug out of a 1940s safety talk topic and is still used today as the cause of accidents. One supervisor, on investigating an accident, immediately stated that his injured employee had not sized up the job and that this had caused the accident. I asked this investigator if he had a master's degree in psychology or any training in that field. He replied that he hadn't. I then asked him how he knew that the employee had not sized up the job. His answer: "I didn't know." So if the investigator had neither the qualifications nor the skill to delve into the injured employee's mind, how did he know that the employee had failed to size up the job? When I asked the investigator what he meant by "sizing up the job," he could not give me an answer. Sizing up the job is neither an immediate nor a basic cause of an accident. Rather, it's a cop-out for a breakdown in the safety management system, which is the true cause of all accidents.

At one organization, I was staggered to find that all the accident investigation reports began with the words "Employee failed to . . ." Without exception, every accident had been blamed on "Employee failed to . . ." This is an absolute cop-out and one of the main reasons why safety can become a fault-finding exercise.

Not Following Procedures

Another big cop-out for safety is that an injured worker is often accused of *not following procedures.* This is a general cop-out and is meaningless. It is fault finding and again points a finger at the employee.

During numerous accident investigations, I have put this statement to the test. I have called for the procedures, examined them, and related them to the actual task that the injured person was doing at the time of the accident. Guess what? In most cases, the procedure was outdated or had been lying in the filing cabinet for years, or the employee saw the procedure only when he first joined the company years ago. The questions to be asked are these:

- When last was the procedure updated?
- Is the procedure still applicable?
- When last was the employee trained on the procedure?
- When last was an observation of the employee carrying out the procedure done?
- Does the procedure still suit the job?

In one particular accident investigation, a work team had been disciplined for apparently not following a procedure. They had been disciplined even though there was no injury. The event was an incident in which there was an exchange of energy—a contact but no loss. The potential for loss was extremely high. When we called for the procedure, it took two days for the supervisor to produce it. On examining it, one of the employees participating in the investigation immediately recognized that the procedure was out of date. The whole method of carrying out the task had been changed some two years previously. So the paradigm of assuming that the injured worker did not follow the procedure is a major cop-out.

As the Columbia Accident Investigation Board (2003) said, "Many accident investigations do not go far enough. They identify the technical cause of the accident, and then connect it to a variant of operator error . . . but this is seldom the entire issue" (23).

Freak Accidents

A number of supervisors have had normal, injury-resulting accidents, but on their investigation form they called them "freak accidents." They may have been unusual accidents but certainly were not *freak* accidents. The only freak accidents really are freak accidents. That means they are unusual, completely beyond our control, and almost acts of Providence. Admittedly, a certain percentage of accidents are caused by natural causes. Tornadoes, flash floods, lightning, earthquakes, and volcanic eruptions are normally difficult to predict and control, and when they occur unexpectedly, they can be referred to as "freak accidents."

I'm sorry to disappoint you, but the accidents occurring in your industry are not freak accidents but are genuine accidents. These occur because of poor management control, inadequate standards or application of standards, and a lack of involvement in the safety system. These accidents should also be investigated to uncover their causes, unless they really are accidents caused by the forces of nature or a similar event. Because the accident is weird or a chance-in-a-million occurrence does not mean it is a freak accident, and it should be investigated thoroughly.

Single Cause—Never!

Most accident investigations uncover a single cause and label this as the main cause of the accident. This is impossible. Accidents are always caused by more than one factor. The principle of multiple causes states that *accidents or loss-producing events are seldom, if ever, the result of a single cause.* This means that accidents have more than one cause. When H. W. Heinrich did his research before 1929, only one cause of the accidents studied was considered, leading to the finding that 88 percent of all accidents are caused by unsafe acts. At the same time of this research, the National Safety Council found that if one looked at both the unsafe acts and the unsafe conditions, they were almost on a par.

People feel comfortable when identifying an unsafe act after an accident. This gives them the satisfaction that the employee "messed up." He therefore committed an unsafe act. Examples include the employee breaching the standing safety procedures; the employee deciding to work unsafe; the employee failing to consider his safety and the safety of others; and the employee having an attitude.

The investigation should continue once an unsafe act is identified. If accidents were investigated correctly, other contributing causes would be sought. What about the physical work environment? Are all the machines guarded? Have there been regular audits of the area? Is a structured safety system in place? The major question I would ask when investigating an accident that lists only unsafe acts is, "Why did the system allow the person to commit these unsafe acts?"

Until accident investigation is used as an effective tool to identify the true causes, it is a waste of time. Poorly investigated accidents are missed opportunities to take positive steps to prevent recurrences. Every accident has multiple causes. Proactive safety means investigating thoroughly and identifying *all* the immediate and root causes of the accident. Then it involves asking *why* each unsafe act was committed. It further involves asking *why* each unsafe condition existed. This is opening a can of worms and delves into the basic causes. Remember, the employee who was injured is only a victim of the safety system failure.

As the Columbia Accident Investigation Board (2003) reported, "Causal factors for accidents that result in severe injuries are multiple and complex, and relate to several levels of responsibility" (23).

Treating the Symptoms

Most accident investigations treat symptoms. Unsafe acts and unsafe conditions are only symptoms of a failure of the safety management system and not the cause. Often patch prevention measures are executed after an accident. These only remove the *symptoms* but do not delve into the root or basic causes. To put safety in its rightful place, the actual root causes should be identified and rectified. If this is not done, similar accidents with similar consequences will recur sooner or later.

Accident Investigation Forms

One of the weakest documents in industry and mining is the accident investigation form. The accident investigation form is the key document in the accident investigation process.

Most supervisors tell me that they hate the paperwork involved in an accident investigation. Unfortunately, I normally give them the bad news, which is if they had prevented the accident in the first place, they wouldn't be bogged down with the paperwork.

Incomplete

Most accident investigation forms are filled out incompletely. After an accident and a subsequent injury, I believe that the injured person deserves a thorough and proper accident inquiry. Part of this is completing all the information on the accident investigation form. In some accident investigations that I have evaluated, even the employee's name had been incorrectly listed on the form. This, once again, shows total disregard for safety and also sends a negative message to the workforce.

Even though there is paperwork involved in an accident investigation, the investigation form is a legal document in most states and should be correctly filled out, giving all the facts and not finding faults. Correctly completed does not necessarily mean that an unsafe act alone is sufficient as the only immediate cause of an accident. If the employee has committed an unsafe act, it should be listed as a *fact* and not a *fault*. Numerous accident investigations list only unsafe acts or perceived unsafe acts of employees. What about the system weaknesses?

Who Should Investigate?

Without doubt, the manager of that department should conduct the accident investigation. Whether the manager is the supervisor, foreman, or team leader, he should investigate the accident. After all, the accident occurred in his area to people under his control. The investigation should be an exercise to identify, to him, where the safety management system failed, resulting in the accident.

In some instances, the safety representative may be appointed as the accident investigator, but this should not free the immediate manager of his responsibilities to investigate the accident.

Delegating this responsibility to the safety coordinator or safety officer is not acceptable. The safety coordinator should only assist the foreman in investigating the accidents, as he is an expert in this field and can be used as a resource.

The extent of the loss should not determine who should carry out the investigation. Past experience has shown that once there is a fatality, multiple fatalities, or great financial loss as a result of an accident, only then does top management get involved. What should determine participation in the accident investigation procedure is the *potential for loss of the event* or *what could have happened under slightly different circumstances.*

It should be remembered that the difference between an incident with no loss and an accident with great loss is only a matter of luck. The high-potential undesired event or incident should receive as much attention as the accident that resulted in severe injury.

Safety Department

Often the safety department is used solely to investigate other managers' accidents. The safety department should be used only as a resource and not to investigate other managers' breakdowns. They can be called in to consult, advise, and assist in the accident investigation process. They should not conduct the entire investigation, unless it is an unusual situation that calls for expert investigation.

Training

One of the weaknesses in the safety profession is that people who have to investigate accidents have not been trained in investigatory techniques. Because most of the safety training is slanted toward legal compliance, accident investigation has slipped by as being nice to have rather than a must-have.

Accident investigation is an opportunity to prevent further accidents. Unless it is done correctly, by using the correct form and the correct people, it is pointless. Improving accident investigations will immediately indicate that the past investigations were flawed.

Follow-Up

The most important part of an accident investigation is the follow-up. Even after the traumatic experience of an accident, follow-up measures are normally poor, ineffective, and not executed. If the investigation has not found the basic causes of the accident, the remedial measures will not be effective and, no matter how many follow-ups there are, will not contribute to prevention of further accidents.

Remedial measures should treat the causes of the accident and not the symptoms. Once the problem is identified, solutions should be posed. These solutions should be converted into management action plans that ask the following:

- What must be done to prevent a recurrence?
- Who must do this action? Which individuals, which departments, which supervisors?
- By what date must each action be completed?

Only by assigning responsibility and authority and by creating accountability can safety work be successfully completed. Deadline dates are important, as future, probable accidents do not wait for rectification. They could occur at any time.

Follow-up action must ensure that the steps that were recommended have in fact been implemented.

There must be rigor in following up after an accident. The accident investigation file should not be closed until all the gaps have been blocked and all the necessary steps taken to prevent a recurrence. The acid test after an accident is, "Will these actions prevent a future, similar accident from occurring?" If the answer is "yes," then pursue them by all means. If the answer is "no," why do it anyway?

Monitor

The implemented remedial measures should now be monitored. Inspections and observations should be carried out to ensure that the remedies that have been put in place are effective. Will they contribute to the prevention of a similar event occurring? This monitoring of the remedial measures will ensure that all the checks and balances are put in place and that they are effective. Some remedial measures may not be effective, and a similar accident could still occur. Prescription of a remedy without correct diagnosis is safety malpractice.

Unsafe Condition

I remember the underground miner who once told me that when he was investigating a serious injury caused by an accident, his supervisor told him that there was no such thing as an unsafe condition. This meant that he was obliged to put down only the unsafe acts and blame the injured employee. In fatal accidents, this is often the case, as the employee is not available to defend himself. Some accident investigation forms have a space to write the description of the accident. One particular form I came across headed the paragraph with "What did the injured person do wrong to cause the accident?"

Accident Investigation Is Two-Sided

In managing people in an industrial environment, there are two sides to the coin. Both the employee and management have their responsibilities. This cannot be one-sided, as then it will clearly contradict the principle that all are responsible for safety at different levels of accountability. If safety is to be improved, then both sides of the coin need to be examined. In an accident investigation, both the employee's actions and management controls should be examined and analyzed and recommendations proposed. One cannot merely focus on one side of the coin, as this does not give a true reflection of the problem. Most accidents follow a number of breakdowns, and the errors cannot simply be explained away, as they are often a series of blunders.

Self-Prompting Guideline

The accident investigation form should be a self-prompting guideline that helps identify both basic and immediate causes of the accident. It should be user friendly and guide the investigator through the process. It should list all the unsafe acts and all the unsafe conditions. Basic causes should also be listed.

The accident investigation form must prompt the investigator to get all the facts, list all the contributing factors, and provide a complete description of the accident and allow for the selection of possible remedies.

The Golden Rule

The golden rule of accident investigation is that *accident investigation is fact-finding and not fault finding.*

In other words, use accident and incident investigation systems and methodology to get to the facts, the root causes of the accident, and don't use it to find fault. Finding fault will not help identify and remove the cause of accidents. Remember, don't link any form of bonuses to any degree of injury or disease. This leads to cover-ups of the true causes of accidents.

Summary

Accident investigation can open a can of worms. It should be fair, look at both sides of the coin, and not be a fault-finding exercise. If safety is truly a priority in any organization, an accident will be viewed as an opportunity to investigate, identify where the systems failed, and set up the necessary control measures to prevent it from happening again.

All incidents that have the potential to cause loss should receive the same attention as accidents and should also be thoroughly investigated in the same way and with the same gusto as accidents.

Remember, the injured worker is only a pawn in the safety management system. Most of the time he is an innocent victim. Placing blame for accidents is really shifting the responsibility from where it belongs, with senior management.

13

Paradigm 12: After an Accident, Write a Procedure

ONCE UPON A TIME, a message must have been sent to all personnel involved in accident investigations. That message said something like this: "To solve all future safety problems and prevent recurrence of accidents, write a procedure." In numerous instances, I have found that a procedure is called for after almost every accident. This is a safety paradigm. Procedures do not necessarily treat the cause of all accidents and are therefore not always needed. A procedure is also a cop-out. Once there is a procedure written and filed in the filing cabinet, one can always point a finger and say, "There was a procedure, but the employee failed to follow it."

Parking Lot Injury

To give an example of how ridiculous this paradigm is, the *parking lot injury* is quoted. An employee parked his car in the parking lot and while walking to his place of work stepped into a slight indentation in the gravel-covered walkway, fell, and landed heavily on his knee. His knee was swollen, and he had to receive medical treatment. Because the employee was treated at a hospital and was on crutches for three days following the injury, the accident was investigated. On reading the accident investigation, the employee was blamed for not looking where he was walking. The remedy given to prevent a recurrence was, "Employee will write a procedure on how to exit his motor vehicle and walk safely to his place of work." Now I ask you, how ridiculous can one get? Writing a procedure after an accident does not necessarily fix the root causes of the accident.

Principle of Definition

In safety, the principle of definition should be applied. The principle states that solutions to safety problems can be found only once the real cause of the problem has been identified. Here, the real cause referred to is the basic underlying cause of the safety problem. What the principle of definition helps one do is to find the root causes rather than wasting time treating the symptoms. If this principle is applied in every accident investigation, the proposed remedies will contribute to the reduction of further accidents occurring.

Treating the Symptoms

Much safety effort, unfortunately, treats only the accident symptoms. The symptoms are normally the unsafe acts or the unsafe conditions. These are closest to the exchange of energy that causes the injury. They are prominent because they are obvious. Since most accidents will involve an unsafe act, this unsafe act is what normally appears most prominently. It is often mistaken as the cause of the accident, whereas it is only a symptom.

"Patch prevention" is another name given to these activities. For example, an employee slips on a puddle of oil in the workshop. The accident is investigated, and immediately the puddle of oil is identified as the cause of the accident and subsequent injury. The oil is cleaned up and the accident investigation file closed. Less than an hour later, the same forklift passes over the same area, stops to load a pallet of goods, and drives off, leaving a similar puddle of oil.

The basic cause in this case was the forklift truck, which was dripping oil. To get to the root cause of the accident, the source of the oil should have been sought. Repairing the forklift truck and following up to ensure that the leak is fixed is good safety management control.

Safety Controls

Safety management controls are what constitute a structured, ongoing safety and health management system. Some of these controls may be in the form of procedures, such as emergency procedures, hot work permits, or confined-space entry procedures. They are a valuable part of a safety system and should not be compromised by writing a procedure for everything, especially after an accident.

The controlling process involves first identifying the hazards and thus the risks. The next step is to identify what must be done to reduce the risks to a level that is as low as possible. This would mean setting up action plans, which might include systems, standards, controls, and procedures. Once these standards have been set and accountability has been delegated to where it belongs, an ongoing follow-up system to measure effectiveness should be instigated. Where possible, regular evaluations and audits of these systems should be carried out and corrective action embarked on to fix any deviations. Ongoing recognition and commendation should be given for compliance with standards in all instances.

Validity

If there is a procedure, it may not necessarily suffice. Some questions to be asked about procedures to determine their validity are as follows:

- When was the procedure compiled?
- Does the procedure still apply?
- When was the employee first trained in the procedure?
- When was he retrained in the procedure?
- Is there written proof of his retraining?
- When last was a structured observation of the execution of the procedure carried out?
- How often is this procedure updated?
- Is this procedure discussed at toolbox meetings?

Safety Observations

To ensure that procedures are being followed, observations should be carried out. These observations could be formal observations that involve following the procedure and the employee's performance step-by-step. They could also be casual observations that would entail observing a portion of the procedure while observing unannounced. Formal observations are when the entire operation is observed and recorded with the employee's prior knowledge. Both methods can be used very successfully.

What about the Conditions?

My favorite subject of safety discipline prompts me to make this statement. Far too often, safety discipline is resorted to as the cure-all of all unsafe acts of workers. It has already been explained in detail that unsafe acts are merely symptoms of a weakness in the management system. These are the symptoms and not the causes of accidents. Fixing these perceived problems and thinking the problem has been solved often lulls the organization into complacency.

In one accident where an employee had been disciplined and sent home for three days without pay, I questioned this decision. I visited the work area in which the employee worked and examined the accident scene. I was amazed to find that the housekeeping in the area was way below standard. Approximately 10 percent of the lighting fixtures were malfunctioning, and I noticed at least three fire extinguishers that had not received monthly inspections. This was not only a legal requirement but also a company safety standard. I then asked the supervisor concerned how he could have taken disciplinary action for one unsafe act while he was responsible for all these unsafe conditions in his area of control. It did not make sense to me. Why, if we are going to discipline, was the supervisor not disciplined for the blatant unsafe conditions? Or does this come into effect only once there is an accident and an injury? If it does, then these decisions are based on emotional, knee-jerk reactions and not good management decision making. Or is it a case of "no blood, no foul?"

14

Paradigm 13: Safety Belongs in the Safety Department

PERHAPS ONE OF SAFETY'S BIGGEST stumbling blocks is that the responsibility for safety has been pushed down to the safety department. As soon as there is a safety issue, it becomes the responsibility of the safety department. Safety belongs with the line management, from the lowest level of management to the chief executive officer. Safety is their function. The safety department should only coordinate the safety activities and not accept responsibility for the entire safety function.

Doug's Story

During a staff meeting, a young safety coordinator stood up and made the statement that his management was "misusing him." When I asked him what he meant by "misusing," he explained that he had received a call-out from the factory at 3:00 A.M. one Saturday morning. I asked him why he had been paged and what the problem was. He replied that when he arrived at the particular division, he asked the supervisor on shift what the problem was. The supervisor took him to an excavation, in the middle of the floor, approximately eight feet deep and about twelve feet in diameter. The safety coordinator asked the supervisor, "What's the problem?" The supervisor then pointed to the unbarricaded excavation and said, "Look, this is a safety issue, somebody could fall in here, and that's why I paged you."

I then asked Doug, "Well, what did you do?" "Oh," he replied, "I went and got temporary hand railings and erected a barricade around the hole." I then turned to him and said, "That's why your management is misusing you. You are doing the safety activities for them, and you have accepted responsibility to fix their safety problems, to rectify their unsafe conditions, and you will also be held accountable for the unsafe acts of the employees. No wonder they are misusing you. You are not functioning as a professional safety practitioner."

Responsibility

Safety is not the responsibility of the safety department. The safety department seldom has the authority to carry the burden of safety responsibility. They can be responsible only for coordinating the activities that constitute the ongoing safety program.

Perhaps it's a traditional thing that safety departments want to "do" safety. Maybe this is what make them feel wanted and needed and gives them job security. Perhaps this is a result of their not understanding the prime functions of safety coordination. Some safety personnel want to be called for every issue relating to safety, even if the issue is only slightly related.

It's an Injury—Call Safety!

When it comes to postaccident activities, the focus traditionally falls on the safety department. I have personally seen safety practitioners go for days without a phone call, page, e-mail, or memorandum. Amazingly, as soon as there is an accident and somebody is injured, the safety practitioner gets called. Why? In some instances the safety department is called in to take care of the injured person. This duty belongs with the paramedics, the hospitals, and doctors, not the safety department. The safety department is supposed to be there to promote the *prevention* of accidental losses, not to treat injured people. Safety departments are traditionally reactive and focus all their attention on the degree of the injury.

I know of one safety practitioner who went around to all divisions, gave them his business card, and told everyone to call him if there were any "safety problems."

Chasing the Numbers

Because safety is traditionally a numbers game and is driven by the number of disabling or lost-time injuries, the safety department seems to feel it is their duty to reduce these numbers. Once there is an injury, the only way the safety department can make a difference is by attempting to convince the doctor that the person isn't as badly injured as he is. You will recall the story of the doctor who was most apologetic to me about a particular injury case. He went into great detail to explain to me how the man could not return to work because of the extent of his injuries. I was actually there on a totally different issue, yet the doctor was most apologetic that he could not send the man back to work. He realized that this would result in a lost-time injury and was apologizing profusely. He was so used to being bullied into releasing workers back to work by safety coordinators and supervisors that he was apologizing that he couldn't release the injured person.

I was personally called in to the doctor's office by a manager who spent an hour trying to convince the doctor that an injured employee could return to work. All this bluff was to avoid having to accept a lost-time injury. When I interviewed the medical staff the next day, I found that they were under an obligation to send injured workers back to work to avoid the department having to record an injury.

What was worse was that these injured employees just sat around in the office and did nothing all day. The company then still professes to be lost-time injury free. The safety department was often caught up in the middle of this sham, but their hands were tied by tradition and the old-fashioned safety culture. I managed to change this by getting a standard approved that stated that only the chief medical officer on site could classify an injury. Needless to say, the chief medical officer also came under a lot of peer pressure to classify injuries as minor injuries nevertheless.

On-Site Hospitals

Some large organizations and isolated mines have on-site hospitals fully equipped and manned around the clock by qualified personnel. What do you think their safety "record" is? Good? You bet it is. "Did the man have to leave the place of work because of the injury? No, he's still on the premises; therefore, it's not a lost time!" This is what gives safety a poor reputation among employees and unions.

When a man is injured, he is injured, and the medical department should determine whether he can return to work, not the safety department. Remember the acid test. If he cannot return to *what he was doing* at the time of the injury, it's a disabling injury anyway.

Leave Safety to the Supervisor

Safety departments should move away from injury management. Leave that to the supervisor. After all, it's the supervisor who is responsible for his employees. The supervisor employed the man, he trained the man, he goes through his annual merit review with the employee, he praises the employee, and he disciplines the employee. Now that the employee is injured, shouldn't he continue to remain involved? After all, whom does the man report to? Not the safety department.

Safety Investigates the Accident

The safety department should not be there to investigate all accidents. The supervisors should be the prime investigators in all cases. Safety personnel should be there as a resource, to assist and guide in the accident investigation. After all, if there is an accident, it indicates a weakness in the management safety system, so, logically, management should investigate. There are certain exceptions, however, where the expertise of the safety department may be needed.

Postaccident Interviews

One of the activities that clearly puts safety in a poor light is when the safety department is involved in postaccident interviews with the injured employee. During these sessions, the employee is normally encouraged to work safer and not to do the "stupid thing" he did that caused him to get hurt. These are pointless exercises. They are carried out traditionally by the safety department, and this is wrong. The supervisor of that particular person, no one else, should carry out the interviews. Safety departments, once again, try to get involved in the day-to-day management of another supervisor's employees. This harms their reputation.

Legal Accountability

Part of the reason why the safety department seems to feel that they are responsible and accountable for safety is that a lot of legal agencies make them responsible for safety. This is why safety does not work and why the United States experiences 4 million on-the-job accidental injuries and 8,500 deaths at work each year. Legislation will never succeed in reducing the accident toll as long as it holds safety departments responsible.

To be effective, the mines and factory inspectors, as well as state safety inspectors, should go to where the action is: the chief executive officer. Where possible, all compliance inspections should be carried out with the most senior manager of that particular area. Sending inspectors out with the safety department creates what I call "the safety charade game." The inspector identifies a deviation from a particular regulation, and the safety department rallies to the company's aid by attempting to convince the inspector that the deviation is minor and, therefore, that the company should not be cited. As David V. MacCollum says, "A hazard is a hazard. Stop the charade, accept the fine and then go and fix it!"

After a fatal accident in which a young maintenance worker was killed, a local inspector blamed the safety department as being "ineffective." He also said they had no authority. I questioned his statement and replied that the safety department should not have authority but that authority should lay with management. He also attacked the department because the firm had no formal safety committee. The irony of the matter was that the one general manager had written a memo to the safety department in reply to their request for such a committee stating that the company did not need a committee.

If safety wants to play this role in defending the company or trying to hide the truth, then they should change their titles. They should become the legal department rather than the safety department. The safety department has no authority to take the necessary actions to ensure that the company conforms to the legal requirements. Why, then, do inspectors do the inspections with them? Would it not be far more effective for the inspector to liaise directly with the plant manager and for the plant manager to nominate a senior *manager* to do the necessary inspections?

I believe this situation has evolved because safety departments have tried to justify their existence by professing superior knowledge of local safety regulations. The managers should know as much if not more about the safety legislation than the safety department.

Policeman

A lot of safety professionals take it on themselves to give direct instructions to employees. I have known safety practitioners who do inspections and actually give direct orders concerning the wearing of personal protective equipment to workers. In some instances, employees are told to clean up their workplace because of poor housekeeping. Employees are often instructed to carry out tasks differently by the safety department.

The safety department is not a police force. They should merely identify deviations from standards and notify the supervisor. Surely it is the supervisor's job to give instructions to his employees. Why does the safety department take on this role? If the safety department wants to be policemen, then they should join the security division and walk around giving instructions and correcting behaviors. By acting as safety policemen, they belittle the profession of safety practitioners.

We Appoint the Wrong People in Safety

One of the major downfalls of the safety profession is that the wrong people are recruited, selected, and appointed. Safety will become important only when organizations realize that it is a profession and must be manned by professionals.

Traditional Appointments

In the past, there was no specific pattern for the appointment of people into the safety profession. If safety is ever to be regarded in its rightful place in the management functions, correct selection, training, and appointment criteria must be compiled and followed to ensure that the right person is appointed.

As a young safety adviser, I used to call on refractory industries and quarries in the Olifantsfontein area of South Africa. What amazed me was that nearly every safety officer I encountered had been disabled in one way or another.

The one particular elderly gentleman who had been placed in charge of safety had an arm missing from the elbow down, and another one, who became a good friend of mine, was blind in one eye. Once I had befriended these gentlemen, I plucked up enough courage to ask one of them how he had gotten into the safety game. The gentleman who had lost an eye proudly told me that "after the accident" his management had called him into the office and appointed him as safety officer.

To justify the appointment, the manager told him that he would be a good "visual aid" for safety, as he was evidence of what could happen if people did not wear their safety glasses while grinding.

No One Else Wants Him!

There are plenty of organizations that have safety practitioners manning the safety department because no one else wanted them. In investigating a number of these cases, I have found that these individuals have been transferred from department to department because they were disliked or disruptive or their supervisor could not manage them efficiently. Eventually, in desperation, they were shoved into the safety department and left to their own devices. No wonder safety has such a poor reputation. This would never happen in engineering or maintenance divisions, yet it happens—and will continue to happen—in safety. No wonder we have the constant treatment of safety as the unwanted stepchild.

First-Aid Certificates

One thing that has always amazed me is the link between first aid and safety. In my opinion, there is no link, whatsoever. These are two different ball games. Because a person has a first-aid certificate or is an emergency medical technician or even a registered nurse does not mean that person is automatically a good safety practitioner.

First aid has nothing to do with safety, except that it prevents the injury from deteriorating. Safety is the prevention of accidental losses, and first aid is postcontact treatment to stabilize the person, patch him up, and get him back to work. Numerous organizations appoint first-aid personnel or semi-qualified first responders as safety people. This is a mistake, as safety management is different, for example, from managing body vital signs and restoring resuscitation while stopping bleeding.

National Safety Surveys

Every year, prominent safety and health magazines conduct safety surveys. The results are interesting and clearly show a chink in the safety armor of the safety profession.

The 1999 national safety survey conducted by *Occupational Hazards* (1999b) revealed the following:

- Lack of management support was in general a serious obstacle. A general consensus was a continuing need to educate management on the benefits and cost savings of a safety program.
- More than 50 percent of the respondents said that their staff was inadequate to meet the program needs.
- The majority of respondents used the term "responsible for safety." (No safety department can be responsible for safety. That is management's job. Safety departments are responsible for safety promotion not safety per se.)

- Nearly 40 percent of the respondents said that compliance with the Occupational Safety and Health Administration (OSHA) was their main safety thrust.
- 39 percent reported that safety was considered the safety department's responsibility.
- Slightly more than 50 percent of the respondents said their safety and health programs were very good or excellent.
- 38 percent judged their safety programs as average.
- Only 7 percent rated their programs as poor. (Would you openly admit that the safety program you were responsible for was poor?)

One particular safety official at a mining company said, "Business priorities are changing in the mining industry. Safety is considered expensive and upper management is not as well educated on health and safety issues, liabilities or compliance" (*Occupational Hazards* 1999b, 30).

White Paper Survey

A white paper survey (Johnson 1998b) had some startling revelations concerning the safety industry. Some of the findings as a result of the survey were the following:

- Only 44 percent of the safety and health professionals used management participation and role modeling to get employees involved and motivated in safety.
- Only 36 percent relied on the actions of supervisors to get employees involved. (Whatever happened to the basic safety fundamental that managers lead the safety thrust?)
- The survey indicated that some of the top safety and health priorities for 2000 were supervisors' safety training, ergonomics, respiratory protection, and behavior-based safety.
- More than 80 percent of the safety personnel indicated that their job responsibilities had expanded during the past five years.
- In only 24 percent of the organizations surveyed did managers and employees trust each other.

Job Satisfaction

- Less than 50 percent of the safety personnel felt a rewarding sense of job satisfaction. This is an improvement on the 33 percent in 1995 and the all-time high of nearly 50 percent in 1997. The question I ask is, if more than 50 percent of safety and health professionals are unhappy at work, how effective are they?

Support

- A large number of safety personnel (66 percent) get management's support through OSHA/Environmental Protection Agency compliance. In other words, they use the law to motivate their management. Second in line was the using of injury and illness statistics, and third was workers' compensation costs. High on the list at number four were concerns for liability that was used to get management's support. Risk assessment rated third from the bottom at 26 percent.
- To involve and motivate employees, most professionals use training and education, followed by rules and policies and ongoing communication in the form of newsletters, bulletins, safety poster boards, and so on. Positive reinforcement fell to number five, and, disappointingly, nearly a quarter (24 percent) still used discipline to get safety involvement and motivation from employees. (This is a waste of time. Safety discipline has never—and will never—work as a safety motivator.)

Job Security

- The survey showed that 15 percent of the professionals are worried about job security, but a quarter of them admitted, "Safety is subordinate to production and other performance goals in their company." (Hence the paradigm that safety *is first.*)

Multiple Functions of Safety

Safety was once a job function of the safety practitioner. Then occupational health was added to his portfolio. Then, because everyone wondered what the safety guys did, they gave him environmental management as well. Thinking that these are the same thing is a paradigm. Safety cannot be watered down and burdened with other functions such as quality and reliability. Safety and health have the same objectives and should stand alone.

Guide, Educate, Train, and Motivate

The function of all safety departments should be to *guide*, *educate*, *train*, and *motivate* all levels of management, workers, and unions in the techniques of accident and disease prevention. This should be a staff function and not have a line responsibility at all. This has been stated in numerous instances by a number of safety authors. Yet safety departments traditionally drift back to try and ward off any citations from legal agencies, whether the company is at fault or not.

Safety Is a Profession

Safety and health engineering and management is a management science and is a profession that doesn't get the recognition and support it deserves.

Safety departments should become far more professional. Their true place is advising management and coordinating the activities of an ongoing safety management program. They cannot improve the safety by accepting the responsibility for safety. They should not directly try to influence behaviors of employees. Only management can do that. The traditional safety we know is a thing of the past. New approaches are needed if safety is to reduce the risk to people at work.

Professional Safety Development

One of the reasons that safety lags behind is because of the lack of professional development of safety personnel. There are a number of people working in the safety profession who got into the profession for a number of reasons. Some were injured in accidents and transferred to the safety department either as punishment, demotion, or for whatever reason. Some entered the safety profession because they were no good at anything else. Instead of being fired, a position was created in the safety department for some individuals.

On the other hand, there are professionals who have embarked on safety management as a career. They have spent at least four years at college studying a degree in occupational safety or industrial hygiene and rate themselves as professionals. A message to anyone in the safety profession is that the safety profession is becoming professional, and formal qualifications and ongoing self-development are necessary to make a success of the career. Safety can no longer be a dumping ground for employees who do not fit in with the system for one reason or the other. It must be recognized as a vital management function and a professional function run by professional, qualified, and competent people.

Interpersonal Relationships

One of the main requirements for a good safety practitioner is excellent interpersonal relationship with others. Unless a person with the right communication skills, approach, and motivation is appointed, constant clashes could occur. Often safety practitioners who have been dumped into the safety department lack these skills and create safety conflict. This, then, turns people against the concept of safety, as they relate it to being constantly forced "into safety compliance." Unless exceptional interpersonal skills exist, safety becomes a forcing rather than guiding, educating, training, and motivating process.

Education and Qualifications

Not everyone can be appointed as a safety practitioner. Because the person has an engineering degree does not mean that he or she can be appointed as a safety practitioner. I believe most management think that safety is merely "common sense" and therefore that anyone with common sense can manage the safety function. This is incorrect. As with all professions, there must be professional training, professional education in occupational safety and health, and the ability to communicate effectively and enthusiastically.

Numerous safety managers have formal qualifications in industrial hygiene, which, although close to safety management, is really totally different. Industrial hygiene studies do not include management functions and principles, communication skills, or interpersonal skills training. The safety profession must be pure and must be clearly defined as the guiding, educating, training, and motivating of all levels of management in the techniques of accident prevention.

Trainee Scheme

One of the many reasons why the safety profession is not very successful is that there is no approved safety trainee or apprenticeship scheme. With most professions and trades, there is some sort of apprenticeship, training scheme, or other program that a young person can follow to eventually be qualified. This includes both academic and practical experience. Because there is no trainee scheme, it leaves the safety profession with a shortage of young safety practitioners. The current situation is that safety practitioners are retiring, leaving the profession, or progressing to other divisions. This leaves openings that can be filled only with nonprofessional people.

What is needed is some form of trainee scheme to qualify young persons both academically and practically in safety management. Unless such a scheme is instituted, the safety profession will continue to recruit unqualified people.

Flavor of the Month

One of the downfalls of having unqualified safety personnel is that they do not understand the basics of safety. Because the very insight into the cause, effect, and control of accidental loss situations is absent, this leaves them clutching at straws. These straws are often in the form of the current "flavor of the month," such as programs, devices, and other novelties, that are purchased and implemented to try to replace a structured, auditable safety system. These are all safety gimmicks.

Gimmicks, pathetic competitions, and circulation of meaningless safety posters and topics are all part of the safety charade. It is almost like jumping onto a horse and landing on the other side. Skilled and experienced safety practitioners recognize that safety is a management function and that formal safety systems must be integrated into the day-to-day management of the organization. They will also recognize that the safety basics must be in place and that these basics must originate from safety standards set, implemented, and monitored by management. Ongoing evaluation and correcting activities will ensure that the basics of the safety system stay in place.

Flavors of the month come and go and make little impact on a structured safety system. What they do, however, is create mockery in the minds of the worker. The worker knows that these amateurish games, novelties, and other pathetic promotion schemes will come and go and that tomorrow it will be "the same old, same old."

Job/Position Description

The job description of a safety practitioner will differ according to industry need and the incumbent's responsibilities. However, the position charter must contain certain proactive safety activities and should not give the person the responsibility for reducing accidents or the injury rate. This is a line management function.

What helps put safety personnel in extremely difficult positions are their job or position descriptions.

Sometimes these job descriptions force responsibility for safety onto the safety personnel. One of the worst examples quoted the following objectives, among others, for the safety manager:

1. To provide safety information and training that will help in achieving a zero-injury rate
2. Reduce the injuries on site by utilizing a well-trained dispensary staff

Just reading these two task functions makes it pretty clear that the safety person was expected to reduce the injury rate to the magical, mythical, and nonachievable figure of zero injuries.

Concerning regulatory compliance, this particular safety manager had to reduce the number and the cost of the citations issued by the legal entity. This meant that he and his staff were well trained to argue the point with the inspector. If the inspector wrote out a citation as serious, it was basically the safety department's function to reduce the seriousness of the citation. To reduce the costs of these citations, the safety staff often flew to the head office of the legal entities to have a meeting, discussions, and negotiations in an effort to "bargain" the costs.

The safety department should be thankful for the citation, which will help them get the hazard eliminated. Sometimes the basics of safety, such as hazard identification and rectification, are neglected partly as a result of these demands placed on the safety department.

Any safety department that has motivated their management to run a structured safety and health management system would openly invite inspectors from either OSHA or the Mine Safety and Health Administration or other government departments to visit their place of work without fear of being cited for hazards, unsafe workplaces, and/or unsafe practices. Those organizations that have embarked on voluntary protection programs are examples of a positive and correct approach to safety. Trying to bargain down the severity and costs of citations causes us to lose sight of the objective of safety. In many cases, although the citations' gravity is reduced, the hazard remains.

Example of a Safety Practitioner's Job Description

Here follows a job description of a safety, health, and environmental (SHE) coordinator clearly defining the proactive SHE role:

- To guide, educate, train, and motivate all levels of management, unions, contractors, and the workforce in the techniques of accident prevention, occupational hygiene, and environmental control in an ongoing effort to

reduce risk to an acceptable level in order to prevent injury and illness to employees, damage to property, and harm to the environment
- To implement and coordinate the company's safety system, including the environmental management system in his or her areas of responsibility and offer support to management, unions, and workers concerning the safety system

Other Duties

The safety practitioner should meet with the manager(s) of his or her areas of responsibility on safety and occupational hygiene matters once a week for feedback and discussion on the state of the system in the respective areas.

He or she must coordinate and ensure the implementation of the safety system in his or her respective departments and/or areas and must be knowledgeable on safety, health, and environmental regulations, policies, and standards.

The practitioner should conduct bimonthly audits using all elements of the safety system and is responsible and accountable for maintaining the safety element files. He or she must advise all levels on safety, health, and environmental regulatory compliance requirements and should be knowledgeable and proficient on all aspects of the safety system.

He or she must work with safety representatives, managers, superintendents, supervisors, and employees to ensure that proactive SHE goals are met; he or she must coordinate and participate as a resource in safety committee meetings and should be an active participant in disabling (lost-time) injury and high-potential near miss investigations.

One of the practitioner's most important tasks is to ensure the completion of monthly risk assessments in his or her areas of responsibility and to follow up on corrective actions. He or she should also coordinate recognition, suggestion, and observation schemes.

One function vital to safety success is the coordination of incident recall sessions. Another is to test and document deviations of critical systems such as lockout, tag-out procedures, permits (confined space and hot work), lifting gear, fall protection, and contractor safety.

The practitioner has many more functions, and most of them, as can be seen, are proactive activities that help maintain management and employee interest in the safety system and that also ensure ongoing checks and balances focused on workplace and behavior risk.

New Safety Approach

One of the latest safety trends is behavior-based safety. This is the new safety approach and is being promoted internationally by psychologists. Their efforts have been tremendously successful, primarily because the safety gurus could not get their act together, have lost sight of the basics, and have virtually handed the safety profession to the behaviorists. Because of the past brainwashing and the ridiculous notion that 98 percent of accidents are caused by the "behaviors of people," behavioral safety is an ideal cop-out that helps create this paradigm.

Management Behavior

Because we wrongly believe that mainly behaviors cause accidents, the behaviorists are now attempting to solve all problems by teaching industry to focus on the behaviors of the worker. I believe that if any behaviors need to be worked on, it is the behaviors of line management, from the chief executive officer all the way down to the team leader. These are the behaviors that are responsible for safety and for accidents. If you recall, effective accident investigation identifies not only the at-risk behaviors but also the basic cause behind the behavior. The behaviorists call these the nonenabled behaviors.

As Dan Petersen (2000) says, "Although current research shows that management is the key, many behavioral approaches focus on employees. This allows management to abdicate its responsibility for safety" (39).

Summary

The state of safety in the new millennium is clearly indicative that the safety profession still has a lot of work to do. Safety is still instilled by discipline and fear of legal citations and the consequent liabilities. Management needs to be educated and convinced of the cost benefits of safety systems. I realized that there is room for improvement when I attended a safety and health conference and noted only one manager present out of a total of forty people.

My challenge to the safety practitioner is to let executive management take ownership of the safety system. My advice to you is, "You are the promoter, the coordinator, the adviser. Your job is to guide, educate, train, and motivate. Remember that safety is *not* your responsibility. Delegate it to where it belongs, with all levels of line management." It's time to change the paradigm that safety belongs in the safety department.

15

Paradigm 14: Safety Is Boring

WITH ALL THESE HOLLOW EXPRESSIONS such as "have a safe day," "have a safe one," "work safely," "think safely," and "safety is first" being used to promote safety, safety *has* become boring. Safety can be the most exciting science and profession in the world. It is common to all and affects us at work and play or while traveling. Thinking that it is boring is a paradigm.

Safety Training Has Become a Drag

I once walked into a training center and was preparing to present an eight-hour course on the systems, standards, and controls program. As I was preparing to open the class, a student entered the room and approached me. He asked me if this was the correct room for the eight-hour safety training course. When I told him it was, he asked if he could quickly return to his truck to fetch a few things. I asked him what he needed. He said he just wanted to go back to his truck to get his pillow. I asked him if he had a problem with sitting for long periods or some other problem, and he replied, "No, I normally sleep through these eight-hour safety training classes."

In most workplaces, safety has become boring and a drag. Safety training has in some instances become almost painful for the candidates having to attend the classes. Safety talks, committee meetings, and coaching sessions consist of going through the safety motions. This sends out a strong message that safety is merely a charade that must be played anyway, so let's get it over with.

Adult education must be fun. If one looks at the constant bombardment of information we receive, one boring lecturer talking in a monotone and occasionally writing something on a whiteboard in front of the class is a sure recipe for safety boredom. Showing ten safety videos in a row is just as bad.

Safety Lacks Enthusiasm

Maybe safety has become boring because the people promoting it are incompetent. The message to safety people who have to present safety training is this: If you are not enthusiastic about your safety training, don't do it. If you are presenting the training to go through the motions, don't do it. This approach has given safety training such a poor reputation that it is obviously a reason why safety has become a bore. Get excited about safety. If you get excited about safety, the people you are training will also get excited about safety. If you apologize at the beginning of the lecture for a drab, boring subject and make excuses for the fact that it is legal training and you *have* to go through it, change your job.

People have approached me in the past with absolute dread for their annual eight-hour safety refresher class. I have had employees say to me, "I hate that day more than anything else." Is this what safety has become? If so, then the presenters of safety topics are part of the problem, and they should become part of the solution.

Attendees Sleep

The biggest problem with safety is that it lacks enthusiasm. Employees are not enthusiastic about safety because safety becomes a disciplinary thing as soon as there is an injury. They are not enthusiastic about safety training because safety training is a bore and a drag. I have seen people pay up to $2,000 to attend safety conferences, only to sit in the back row and sleep through the entire lecture. Is this the result of the belief that safety always has to be drab and boring? Safety education is a two-sided coin. Either the lecturer is boring, or the attendee was out on the town the night before. What a waste of an employer's money for people to sit and sleep through safety lectures.

Safety Training Can Be Fun

Safety can be exciting, enthusing, dynamic, and the most interesting subject to discuss, lecture on, or listen to. The approach is important. The people involved in communicating the safety message or promoting the safety mission are what make the difference.

Get Trained

My advice to anyone who has to present a safety lecture, talk, or training class is to go out and get trained in how to lecture. Join a club such as Toastmasters International. Make sure that you use visual aids. Make sure that you know what you are talking about. Get people to participate. Don't just show videos. Make safety exciting so that your class will come up to you afterward and say, "You know, safety is exciting and can be fun." Remember to let the students do most of the talking.

Visual Aids

Safety training calls for the use of visual aids. Few safety lecturers use visual aids. Why? Since the invention of the video and the VCR, some safety people seem to think that they can walk into a lecture room, lay down the rules, turn on the VCR, leave the room, and go and drink coffee in their office. This is the worst method of safety training. It is the least effective, and if you interview employees, they will tell you that it's a bore.

Visual aids can include full-color presentations made on computer software programs such as PowerPoint™. These presentations can include movement, sound, and even video clips. What about group exercises in the classroom? Frequent breaks are a must. Incident recall sessions should be held at every safety meeting, training session, or training class. These are the most interesting learning sessions that the students could be exposed to. Color slides and case studies can help convert the intangible concept of safety to tangible pictures that students can associate with. Students can participate by filling in handouts. Role-plays are ideal for safety.

Role-play

I teach accident investigation by going through a role-play. I assure you that it is a tremendous eye-opener for the students. At the end of an eight-hour class, they have said to me that they've never enjoyed training so much in all their lives. They all comment on how fast the time passed. They comment that never before have they been exposed to such fun learning.

Challenge

My challenge to you is this. If somebody falls asleep in your lecture, then you are not doing a good job. I have sat through safety training courses for four hours without a break. The average person's span of concentration is twenty seconds. Look at your television screen. How long is the average advertisement—a maximum of thirty seconds? Yet, in safety, we expect people to concentrate, pay attention, and take in information for four hours nonstop. That's why safety has become boring and the paradigm has been created. We have not yet caught up to the twenty-first century in safety communication, lecturing, and training.

I have challenged my students over the past thirty years that if any of them could fall asleep in my class, I would buy them a six-pack. I have yet to buy a six-pack. Some of the basic rules we use to ensure that safety training is fun are first to make it fun. Another is that there will be no wrong answers in the class. This encourages participation. Yet another is that the class is so structured that every fifty-five minutes there is a five-minute break. One of the students is appointed as the timekeeper and given a referee's whistle that he blows at break times. The last one seated in the class after a break receives the position of "honorary timekeeper."

Ground Rules

To keep up with the times, the ground rules at the beginning of the class must include a request for all cell and mobile phones to be set to the silent mode and that no calls are to be made or received during the lecture. Sounds crazy, but I have seen this happen in a number of safety seminars and courses. The students and I agree on these ground rules at the beginning of each session. Another agreement is that if a cell phone rings, the culprit must stand up and sing the National Anthem. This also goes for latecomers after a break. I once had a Filipino in my class who, after his first late arrival after the break, sang and then received a standing ovation from the class. He was good. He also came late after every other break. He told me afterward that he had to finish the other verses.

Getting Their Attention

While working with miners who had been indoctrinated by a fifty-year-old safety culture that included the paradigms that "safety is boring" and that "employees cause accidents," we had to make safety training enjoyable and beneficial.

Confronted with very poor job observations of critical tasks, we devised a good example for a two-hour training session to make safety exciting.

During the first hour of the lecture, we revisit the steps necessary to write a written safe work procedure. These are the listing of the occupations, the listing of the tasks within the occupation, and the critical task risk assessment process. Once the critical tasks have been identified, the top 20% are analyzed by looking at the steps of a specific task, the risks within each step, and the necessary controls to mitigate the risks.

Observing a worker on the job following the procedure, step-by-step, is crucial to the process, and training in this methodology is important.

Creating a Disturbance

While the first one-hour session is in progress, a colleague enters the room and executes certain prearranged activities. First, he bangs the door, then he moves to the corner of the room and proceeds to unpack a filing cabinet standing in the corner. He then moves over to the thermostat and adjusts the temperature of the air-conditioning unit. Next, he moves to the pencil sharpener, where he proceeds to sharpen a pencil for one minute. During this time, he is making as much noise as he can, even though the lecturer is ignoring him. He finally walks right across the front of the classroom and exits the room.

After the break, the students are handed a paper containing a diagram of the room and a sketch outline of a person. They are asked to fill in their observations as to the person and the activities that went on during the last lecture. Amazingly, some students don't even know there was another person in the room, almost creating havoc during the lecture. Some students actually score pretty well. Most of them are so focused on the lecture that they fail to observe deviations from the normal, which is important in job observation.

At one organization during this particular exercise, one of the students actually got up, approached the person carrying out the activities, and asked him to please leave the room, as he was disturbing the class. This was a great reason for a good laugh during the debriefing.

The importance of the exercise, which is pointed out to the students, is that if they want to do an observation, they must be aware of the circumstances around them and must focus on what the employees are actually doing, especially if they are involved in executing critical tasks.

The Eggs and I

After the exercise, we hand out a six-page written safe work procedure. ("Job Safety Analysis" or "Job Safe Procedure," and so on). This work procedure is for the critical task of "scrambling two eggs." The job classification is "Chef Level 1." Four pages list the forty-eight steps on how two eggs must be scrambled in the correct sequence, taking all the necessary precautions. The last two pages are diagrams of how the equipment and materials must be laid out on the table. They include outlines for the placing of the knife, fork, and spoon as well as the salt and pepper containers, pots, and so on (good housekeeping rules).

Five-minute task training is then given to the entire class. The forty-eight-step procedure is explained to them in detail. Some actually shake their heads and say that the procedure is ridiculous. "Forty-eight steps to merely scramble two eggs, crazy," they said.

Volunteer

Once the class agrees that they understand the procedure and there is clarity on all the steps, correct methods, and sequence, we call for a volunteer from the group. He or she is brought to the front of the classroom. A table has already been positioned on the lecture platform. He or she is then instructed to carry out the "critical task" by following the procedure step-by-step. Located in the corner of the classroom, on the floor, are two cardboard boxes labeled "materials" and "equipment," respectively, that contain the ingredients and appliances for the task at hand.

Students at this stage normally want to mime the action. They have been so used to being bored to tears by safety training that they think of this as a miming exercise. Most times we have to point to the materials stacked on the floor and say, "We really want you to do this." In amazement, they say, "Do you want me to scramble eggs here in the classroom?" We reply, "Yes, we want you to follow the procedure and do the critical task while the rest of the class observes you doing the task step-by-step." After all, this is the objective of a planned job observation of a critical task.

The volunteer then reads step 1, which is, "Lay out the material on the table as per attached sketch." Immediately, he walks to the box standing on the floor, bends over, and picks it up to carry it back to the table. In most instances, the wrong lifting technique is used. The other students have been instructed to observe each step and write their comments, either positive or negative, in the column provided opposite each step on the written safe work procedures.

Observation

Often, we stop the class here and ask how many students noticed the deviation from the procedure. Many miss the first mistake. Some pick up the incorrect lifting technique. When asked what observations they have made, it is normally "To teach the employee how to lift correctly." We then explain to them that there were multiple causes that had potential for injury in this step. The other cause was that the boxes had been stacked on the *floor* and not at waist level on a suitable rack or other table. The person had to bend down awkwardly because the material was on the floor to start with. This immediately creates an interest in the observation process.

Keeping Their Attention

Going through the forty-eight steps and observing what the "chef" is doing keeps students on the edge of their seat for the next hour. This forty-eight-step critical task has thirty-two deviations in it. One of the steps that is missing in the procedure is the instruction to plug the hot plate in to the outlet. Some volunteers are so intent on following the procedure that they actually pour the egg mixture into the frying pan after having turned on the hot plate as instructed by the procedure, all this while the hot plate has not yet been plugged in.

The point we make here is that when people follow a procedure, they could create an unsafe act or condition as a result of the procedure. In this case, the procedure did not instruct the person to plug in the hot plate, which is a critical step. Poor or incorrect procedures often cause accidents.

It is important to make safety training exciting, and the "egg and I" example shows how it can be done.

Summary

Safety can be exciting if you make it exciting. By becoming enthusiastic, you will make people around you enthusiastic.

Put fun into safety training. Use visual aids, get people to participate in group exercises, and utilize role playing and attention getters. Finally—and this is very important—get trained in how to lecture.

If we want to change this myth of safety being boring, start putting excitement into safety training. Whether this safety training is a five-minute talk or an eight-hour annual refresher training required by law, make it fun. By doing so, you will be helping to change the paradigm that safety is boring.

16

Paradigm 15: Safety and Security Are the Same

Around the world, safety and security seem to have become synonymous. This is unfortunate, as they are totally different professions. That they can be coupled as one and the same profession and viewed as similar professions is a paradigm.

Catch People Doing Wrong

The prime function of security, the site security officers, and guards is to catch people doing things wrong. Their intent is to protect against theft, intrusion of privacy, and consolidation of intellectual property. They are also constantly on the outlook for unauthorized people entering premises. They monitor the flow of vehicles, machinery, and equipment into and out of a work area. Make no mistake—security is vital and plays an important role in industry, commerce, and mines.

There *is* a place for security in most key industries. Key point industries are those that are vital to the functioning of the community or the nation. They can include power stations, hospitals, law courts, police stations, water pumping installations, and so on. These need to be protected, as they are targets for sabotage from unruly parties, disgruntled workers, or other adverse organizations.

It is, however, not the same function as the occupational safety and health management function and should never be combined with or confused with safety.

The main objective of security is to catch people doing things wrong.

Catch People Doing Right

Safety's prime function, in comparison to security, is to catch people doing things right. It is a motivational science. It involves guiding, educating, training, and motivating people to contribute their efforts to the prevention of undesired events. It differs from security in that it involves more work with the management structure and approach to promote a safe and healthy workplace.

Line Function versus Staff Function

Security is normally a line function. Security is directly in line in the organizational structure. The security manager normally reports to a one-up manager who then reports to another one-up manager. Security gives feedback up the line and receives instructions down from the line.

Safety, on the other hand, is a staff function. This means that the safety department does not report directly in the chain of command but reports in an advisory capacity. Safety personnel are technically "in-company consultants." They advise, guide, and assist management in maintaining a safe workplace, whereas security personnel police the workplace to ensure a crime-free area.

Different Skills

The two functions are distinctly different. Perhaps the only common area is that both safety and security can save the organization a lot of money. Security prevents *intentional* losses to the organization. Safety prevents *unintentional* losses caused by accidents. If safety is to be improved, a compromise should not be made by combining the functions of safety and security. Safety should not fall under the security department and vice versa. Both professions need their own playing fields.

In evaluating job descriptions of the safety adviser and the security officer, there are major differences. Often, to save money, management combines these functions, which is not the ideal situation. Don't get me wrong; sometimes it works, but often it doesn't. Preferably, security should be kept separate from safety. Different skills are needed. A different temperament is needed for safety in comparison with security.

Security has its own professional organizations, and so does safety. Both are entirely different with different aims and objectives. They shouldn't be mixed, as safety efforts will suffer.

Add-On

It's apparent that the powers to be think that safety and health practitioners do not have enough work to do, and as mentioned before, have now added environmental control to the job burden and title. The latest add-on is still coming. It's quality management. Soon the average safety person will be called "Manager: Safety, Health, Environment, and Quality." In some countries, this is already the case. What's next? "Reliability?" Why not. The safety guys have endless capacity. Why not add this to their portfolios?

Safety has always been the dumping ground for other functions, and one of the classic roles that one safety department landed itself was the handling of all the company's film developing and printing needs. Need a film developed? Just drop it at the safety office, and they will take care of it for you. I once worked with an unfortunate safety department that was responsible for all the signage of the factory as well as the telephone connections and arrangements. The same department had never attended a seminar on safety for more than ten years, and the safety superintendent had been in the same position for twenty-one years. Yet they were still used to perform other non–safety-related tasks.

17

Paradigm 16: There Is Honesty and Integrity in Reporting

I HAVE ALREADY DISCUSSED THE LACK OF HONESTY when it comes to the reporting of injuries. The safety cover-up is a major movement to try to keep the injury figures as low as possible—to create a pretense that safety is first within the organization.

Rachel Hawes (1999) has written, "A former senior ESSO executive has accused the company of covering up workplace accidents, fudging figures on injuries and burying safety reports" (5). If a major organization like ESSO is accused of fudging the figures, what about other major organizations?

The Uncertainty Principle

Many safety professionals say it is this single-minded focus and attention on statistics that often skews employees' perceptions.

They also agree that once the measure "injuries" becomes a goal, employees, managers, and safety departments feel pressure to reduce the number of injuries, not necessarily the conditions and behaviors that lead to injuries.

When referring to statistical manipulation, most agree that statistical manipulation may produce swift initial results, but it is unable to sustain improvement because it is based on the premise that actions are not taken until injuries occur.

In supporting the paradigm that safety reporting is honest, most safety professionals are hesitant to openly talk about some practices they say may also include inappropriate practices such as paying for prescription medications from petty cash funds or out of pocket to avoid reporting. Some ongoing cover-ups require employees to take vacation to cover lost time, hiding or the destroying of injury reports, and ignoring (and thus condoning) an employee's efforts to hide an injury.

Is a Hernia Not an Injury?

During the safety orientation training of contractors at a major construction site, a student confronted me. He was sitting in the front row, and when I made the statement that the policy of the site was that no one would get injured, he asked a question: "What about me?" He then related his story. While hoisting a heavy beam, he had grabbed the end of the beam to stop it from swinging. In doing so, he had overexerted himself and caused a hernia. He was immediately sent to the hospital, treated, and put on light duty.

To ensure that no injury was recorded on this particular site, which had the objective "No one gets injured on this site," he was immediately transferred to another site, some seventy miles away. Only on his return to the original site did the truth surface that he had been on light duty for two weeks, was under medication, and could not carry out his normal duties. This was obviously a case of fudging the figures and transferring injured employees out of the work site to retain the injury-free record.

E's Story (As Told to the Class)

"I am a twenty-five-year-old woman with the body of a forty-year-old," she said during the interview.

"I was moving fifty-five-gallon drums in a storage area. My equipment operator was in a hurry and didn't want to do the job. He was very resentful that he got stuck with the job. He started rushing the job, and he was going too fast, dropping things. It was a hot day.

"When the forklift with the attachment came to pick up the drum, it had to be guided in, and I had to make sure it had hold of the drum before being picked up. We had to push the fifty-five-gallon drum into the hold and keep on pushing it to get the forklift to pick it up. He would come in, but he kept hitting the drums too hard when I pushed them. I was getting more and more cautious.

"Then he came in to grab a load of drums off the pallet. There were only two drums on the pallet. It was an older pallet, and when he came in, he came in too fast and hit the drum. I was trying to tell him to stop, but I didn't realize that my foot had got caught. He kept coming even though I was yelling and screaming for him to stop. He kept coming, and there was a piece of equipment behind me, so I couldn't back up, I couldn't get out of the way.

"He ended up smashing into me with the drum and basically forcing it over both of my feet. The force of the impact hyperextended my one knee, broke one foot, and pulled muscles in the other foot. Later we found that I had a back injury from hitting the piece of equipment that was behind me.

"Well, anyway, he was an hourly worker, so he didn't really get into trouble at all. I had to fill in an accident report, and it had to be signed off by the safety department. I had to go to the hospital that day with a broken foot and everything else.

"When I came back from the hospital, I was asked (I was supposed to take two weeks off) no questions. Two weeks off, which was what the doctor originally said. When I came back, the safety guy said to me, 'Well, if you take time off, it's a lost-time accident. We will lose our bonuses, and the safety department will lose their bonuses.' The bonuses were being paid out soon. [What were the bonuses worth?] It depends on how much you make. Mine was about $800. The supervisor's was probably about two grand. We only got bonuses once every two years. This is when the price of the product was still high and everybody had done a real good job. Our division was one of the prize properties at the time.

"So, anyway, he asked me if I could come to work, and they would give me a light-duty job. I wouldn't have to leave the office space. It's like working. Well, I didn't know what I was supposed to do, you know. The basic impression I got was that I had to go ahead and do the light duty. I was brand new there as a salaried individual. So he calls the doctor back and asks the doctor if I am put on light duty, will he release me? Well, the doctor was the company doctor, and of course he was going to say 'yes,' you know.

"So they released me, but it didn't turn out that I had a light-duty job. I ended getting stuck with my same responsibilities. I had to do the work. I had to put my boots on, even with my broken and injured foot. I was in physical therapy for my injuries, but I did not have the time to do all the exercises. My injuries did not heal correctly, and I am still having problems with my foot, my knee, and my back. We never had to report a lost-time accident. The safety record was intact. They already got their bonuses, and to this day they still report that they hadn't had a lost-time accident" [*injury*].

Porta Cabin

One organization I encountered believed that if a worker was still "at work" or on the property after an accidental injury, the injury was not a lost-time injury, which was their main measurement of safety. Consequently, injured workers were collected in the morning at their homes by the company driver and transported to work. These unfortunate accident victims then spent the rest of the work shift sitting in a specially erected temporary work office called the "*porta* cabin." This was a local joke to all employees but is the absolute truth. I further discovered that other organizations in the same area also had *porta* cabins full of injured employees. Their lost-time injury rate remained very low. I wonder why.

Production Not Affected

On July 31, 1999, a methane gas explosion killed eighteen miners at the world's deepest gold mine, Nponeng, in Carletonville, South Africa, eighty kilometers southwest of Johannesburg. The Reuters report on the Internet stated, "The National Union of Mineworkers, which represents the majority of gold miners, was quick to condemn AngloGold safety procedures." The report ended by stating, "AngloGold said it did not anticipate that production at the mine would be affected by the accident." This statement obviously puts production as a priority over the deaths of the eighteen workers. Safety first? Never. Where is the integrity of safety here?

Safety Incentives

Safety incentives help drive injuries underground. The greater the incentive, the greater the degree of dishonesty. One safety incentive is a jackpot scheme, where a scratch-off game card is used and employees who are accident free (should read *injury free*) and meet specific safety criteria receive such a game card. The points are accumulated, coupled to dollars, and traded in for brand-name merchandise. The producers claim that "hundreds of companies had used the program to improve their safety records."

What does improving their safety records mean? I believe that it improves the employees' ability to hide injuries. One statement offered by the product producers is, "Lost-workday cases (LWCs) were reduced by 100 percent, and off-the-job LWCs decreased 75 percent. The company has gone 150,000 man-hours without an accident and remained injury free in 1994."

When I read these claims, I immediately get suspicious and also question any organization going 150,000 man-hours without having "accidents." Do they mean 150,000 man-hours without somebody being *injured* as a result of an accident, or 150,000 man-hours without any undesired event that result in any form of accidental loss? I also question the promoters of these products. If they can't get basic safety management language correct, how can they improve safety at the workplace?

Scrutiny

The Occupational Safety and Health Administration (OSHA) classifies safety incentives into two categories: the *traditional* incentive scheme and the *nontraditional* scheme. Traditional incentive schemes have such a reputation of forcing workers to not report injuries and illnesses that OSHA has had the entire concept under scrutiny for a long time. OSHA felt so strongly about the negative connotations of a traditional incentive program that they were on the verge of drafting proposals for their Voluntary Protection Program participants. *Nontraditional* schemes are based on the safety and health efforts made by the organization rather than the consequence of accidents.

Numerous organizations and promotional companies in the marketplace are making a fortune selling so-called safety incentive schemes. These are in the form of safety bingo, safety cards, and other frills that do nothing but encourage workers to become clever about hiding and not reporting injuries.

Monetary Awards

While speaking at a safety and health conference, Bryan Wollam told the delegates, "Never, give money for any safety reward whatsoever!" Bryan was certainly right. Money has often been quoted as the root of all evil, and if one listens to the news, it concerns murder, robbery, extortion, and other vices, all in an effort to get more money. Why should safety be different? Pay a man to be accident free, and he will be accident free. Money is a great motivator. Whether the money will encourage him to work safer or whether the money will change his attitude (which I doubt) seems to be irrelevant. Money in the form of a safety incentive will reduce the number of injuries. They will be driven underground.

Some organizations have a policy of "No injuries under any circumstances," so what do employees do? They do their best to hide injuries.

These incentive schemes constantly refer to working without having an accident. The correct terminology is working without having *an injury caused by an accident.* In the new millennium, we are still using terminology that does not make sense and is meaningless.

I recently attended a safety awards presentation dinner where individuals were given a check for being injury free. If one reexamines the true cause of an accident that injures a worker, it will appear obvious that injury free could be translated as being lucky. If workers committed one unsafe act or created one unsafe condition during the period in question, we would be rewarding them for that behavior. I challenge you to ask one of those workers what he or she did to prevent being injured. Not one will be able to list five specific activities that they did or actions they took to be accident free (injury free). In fact, you will find they were not accident free. Because of luck factors 1, 2, and 3, they were injury free.

My advice on safety incentives is to not give any reward for a worker having worked any period of time without suffering an injury. My second piece of advice is to never, ever, give cash as an incentive. Cash is slipped into the pocket; it's used to buy the groceries or fill up the tank with gasoline and disappears with the normal day-to-day expenses.

The best type of award to give is an item, such as a jacket, shirt, pen, necktie, or some other item, that will not disappear two days after it is awarded. I was personally awarded numerous safety pens during my safety career and eventually had a collection of forty-eight engraved and personalized pens. These were reminders of the particular safety achievements, and I treasured them and kept them. The same would not have happened with money.

Safety System

An incentive scheme does not replace a safety system or a structured safety management program. An incentive scheme should form part of a structured safety system and should be based on upstream safety measurements rather than measurements of failure, such as injuries and accidents.

The best criterion for any safety award is the answer to the following question: "What have you done to promote safety during the past twelve months?" The answers to this question should be in terms of, for example, safety inspections conducted, risk assessments carried out, safety training programs attended, safety talks presented, and frequency and accuracy of checking ladders, fire extinguishers, lifting equipment, and so on. All of the more than seventy elements of a structured safety program have measurable activities. Measuring these activities and evaluating them is the only criteria on which to base safety incentives.

I once ran a national "Safety and Health Representative of the Year" competition. To choose the winners of this competition, the finalists were interviewed by the executive management committee and had to present an entire portfolio of activities and reasons why they should be nominated as the "Safety Representative of the Year." These were all upstream activities. The quality of the person's monthly inspections was evaluated, and the way in which he promoted safety in his workplace was scored. The tenacity that he showed in promoting safety was also taken into consideration. These are the positives for safety incentives. If the philosophy of the luck factors is studied in detail, it will be clear that the outcome of any undesired event is fortuitous. That means it is based on luck. If awards are then given to the effects and results of these events, they are indirectly being given for luck.

Peer Pressure

Traditional incentive schemes create tremendous peer pressure that leads to dishonesty in reporting injuries. Should incentives be given for the injury-free performance of groups or work teams, one person's injury could adversely affect the entire team. I've asked this question a number of times: "Would you like to be the one that causes your team to lose their $400 bonus for being injury free?" Everyone answers, "No!"

Safety Blackmail

This type of setup is safety blackmail, which is being encouraged and supported by management. All management focuses on are the downstream measurements that regretfully are the only gauges of safety performance that most people know and use. By dangling the carrot of financial incentives for being injury free, the most powerful forces are put into action in the form of peer pressure, fear, and dishonesty to achieve the prize.

Nothing is achieved by these incentives. Do these incentive programs improve the housekeeping? Do they ensure that an ergonomic survey will be carried out? Do they ensure that executive management takes the lead in the safety management program? The answer to all these questions is no. They do nothing to promote a healthy workplace. They attempt to focus the employees' attitudes to work safely, but who says that is the cause of the accidents and subsequent injuries?

What about the property damage accidents? What about the near misses? There are at least a thousand near misses that occur before an injury-causing accident occurs. If incentive schemes are focused on the injury part, what about the near misses? Does an incentive scheme reduce the base of this accident ratio triangle? No. Therefore, they are *useless.*

Workers do not come to work in the morning planning to be injured. The fact they are injured does not necessarily mean that they have worked unsafe. Incentives to be injury free do not create safe attitudes or a safe and healthier workplace.

Quotable Quote

Here are some quotes from users of traditional incentive schemes, which clearly indicate that safety ignorance is bliss:

- Our organization has a dinner each month the organization goes accident free.
- Employees at different divisions receive jackets, bags, and so on for each month of accident-free performance.
- Employees receive merchandise for each year they go without an accident.

These quotes clearly indicate that we are not talking about *accidents* in the first place. They are really talking about *injuries*. That company is not going to experience a day without having an accident, in the true meaning of that word. Employees may get jackets and merchandise, but their organization is not yet in control of safety, and a thorough baseline safety and health audit would find plenty of worms in the woodwork. Plenty of risks probably exist in the plant.

Traditional safety incentives do not put safety first. They put a reduction of the numbers first. A reduction of the numbers by any means, even if it means bribing the employees with money, jackets, and other merchandise, is the order of the day, and it is a sham. Believing that there is honesty in reporting injuries caused by accidents at work, especially where traditional incentive schemes are in operation, is a paradigm.

18

Paradigm 17: Ridiculous Safety Prevents Accidents

I RECENTLY ATTENDED A MAJOR CONFERENCE attended by safety and health professionals from around the country. Some of the stories that I came away with deserve repeating here. All pertain to this paradigm, which is that ridiculous safety gimmicks and campaigns prevent accidents and actually improve safety in the workplace.

The Rubber Mallet and the Candy Bar

Apparently, one organization was experiencing a number of injury-producing accidents and decided to take some action. The safety department carried out inspections of all divisions, and anyone found contravening a safety rule was playfully tapped on the head with a rubber mallet by one of the safety guys. Other workers, found to be complying with safety rules, were given a candy bar by another safety person.

I can just picture the safety clown prancing around the workshop tapping people gently on the head with this large rubber mallet. He would say, "You were naughty, you didn't work safely." And the other safety clown tiptoeing around the workshop like a ballerina on a mission, patting rough, tough, and ready artisans on the back and giving them a candy bar and saying, "Thank you for working safely. Here's a candy bar for you."

The more I thought about this scenario, the more ridiculous it appeared. Can you imagine what the workers thought of this fiasco? What stories would they tell their buddies while sitting in the bar over a beer that night? What sort of a reputation does this give safety? Sure they were attempting to apply some very basic management motivational theories. Unfortunately, they boxed it up. As long as silly promotional pranks such as these form part of the safety scenario, safety will be regarded as an absolute waste of time. Not only did this effort, in my opinion, belittle the intent of safety, but it made safety appear to be childish and ridiculous.

The Cheerleader Squad

Another story that I must relate concerning safety promotion is that of a company that called in the cheerleaders. Apparently, they were experiencing a high injury rate as a result of accidents, and once again the safety department was instructed to "do something about it."

The woman relating the story was the safety coordinator at that plant, and she told us that they had recruited a team of cheerleaders. In full regalia, these cheerleaders then pounced on each department and posed for photographs with the operators, mechanics, and other workers in the field. The resultant photograph pictured smiling, happy cheerleaders and surly workers posing together.

I cannot find the connection between the cheerleaders and safety, but apparently this was termed a "safety campaign." The visit by the cheerleaders was, I suppose, intended to cheer up the workers or change their attitude or something. I still don't know to this day. What I do know, however, is that it seemed to be a ridiculous thing to do. How could cheerleaders visiting the workplace possibly improve the safety? Wouldn't it have been far more effective for management to have visited the workplace? What was the purpose of taking photographs posing with the cheerleaders? If this is safety promotion, then it has failed. What it has done, however, is make safety and safety promotion schemes appear more ridiculous in the eyes of the average worker.

Safety Signs and Motivational Banners

Safety signs, motivational banners, and safety publicity boards are a joke. They send a weak message to employees and are a waste of money. Their messages are outdated and ridiculous.

Coming across an advertisement for motivational banners, these were some of the slogans offered:

"Safety has no quitting time."
"Take time to work safely."

"Three causes of accidents: I didn't think, I didn't ask, I didn't know."
"Forklift truck drivers must think safety."
"Safety is a team sport; play it safe."
"Forklift drivers, does safety have your attention?"
"Work safely today."
"Be careful, be safe, be aware."
"Our goal: no accidents."
"Safety and quality are the top priorities at [your company's name here]."
"Pay attention to safety; don't get carried away by overconfidence."
"Let's make this year accident free."
"Think safe, work safe."
"Zero injuries and beyond."
"Safety is our priority."

Is this up-to-date advertising? I believe these slogans are meaningless and pitiful. They're an absolute joke. Hanging banners like these all over the workplace is not promoting safety and is certainly a false indication that the organization cares for the safety of its workers. These banners are normally erected and left to hang for years in the workplace while they gather dust and become part of the wallpaper. I once interviewed a group of employees in a maintenance area and asked them what banners were displayed in their work area. None of them could recall what banners were displayed. In reality, there were four massive banners that had been strung up many years ago. They had faded into the background, and the workers didn't even know they existed.

Hair: The Safety Version

What is it that links the hair on a person's head to safety achievements? I find it strange that one particular supervisor challenged his crew that he would have all his hair shaved off if they completed a certain number of shifts without having an injury. This they did, and his hair was shaved. Great fun for all, and we think we have improved the safety of that crew.

What actually happened was that no employee wanted to do anything to spoil the chances of having their supervisor's hair shaved off completely as a result of a challenge. What a wonderful incentive to drive injuries underground. Most of those workers would have lied, cheated, and stolen to have seen their supervisor walk around with no hair on his head as a result of their teamwork. This is part of ridiculous safety.

A safety coordinator also challenged his division in the same way. He said to them, "If you achieve 500,000 man-hours without a reportable injury, I will shave off every hair on my head."

Guess what? The department achieved their target, and the safety coordinator walked around with a bald head for a few weeks. This is not safety. This is ridiculous safety in its truest form. These gimmicks and antics are not effective approaches to the guidance, education, training, and motivation of all to take safety as a value. They give safety a bad name.

The Old versus the New

One of the reasons for the existence of safety's paradigms is that many organizations are locked into the old safety mind-set. The old theory was that it was assumed that at any given time, workers make a choice to work safe or not. The new theory, however, assumes that accidents are defects in the total system and that people are only part of the system.

The old paradigm assumed that only workers' behavior caused accidents, while the new theory is that all accidents have multiple causes that include both immediate and root causes. These are sometimes referred to as the physical and the organizational causes.

The old school of thought in safety uses the injury rate as the main safety measurement and indicator, whereas the new, enlightened thinking is that there are numerous upstream measures that measure safety control and effort rather than (lack of safety) consequence. Some of these include the quality of accident investigations, committee structures, near miss reporting, training achievements, management and employee involvement, observations recorded, safety standards, and policies and procedures.

Summary

Ridiculous safety stems from the myth that the main way to prevent work accidents is to persuade and appeal to workers to work safely. Modern, out-of-the-safety-box thinking is that workers should be trained, motivated, and encouraged to suggest ways of improving the workplace and procedures to reduce accidents.

Thinking that ridiculous antics are a substitute for a risk-based structured safety system and that they will improve the safety of a workplace is a paradigm.

19

Paradigm 18: Legal Compliance Means a Safe Workplace

THERE SEEMS TO BE A PERCEPTION prevailing in the workplace that if all legal requirements are met, the workplace is safe. This is another safety myth that must be changed.

Legal Compliance is Basic

Most safety legislation covers the basics of safety. These basics of safety are very generic and are not specific to the particular industry. Sometimes the legislation is cumbersome and detracts from the core safety requirements of a particular workplace. Modifying and changing legislation to keep up with changing trends is a long, tedious, and drawn-out process. By the time legislation has been passed, thousands of accidents have already occurred. By conforming to legislation, one does satisfy the judiciary, but does that make for a safe workplace?

Injury and Health Based

Most legislation is based on the prevention of occupational injuries and diseases. As already discussed, these constitute less than 1 percent of the total accidents that could be occurring in a workplace. Very seldom does legislation cover property damage accidents, accidents that interrupt the business process, or near misses with high potential. In some instances, legislation does require the reporting of equipment and machinery failures. This is focusing in the right direction.

Safety and Health Department Responsible

The greatest mistake that any safety legislation can make is to hold the safety personnel responsible and liable for certain aspects of safety violations. This immediately puts safety in second position. Has legislation perhaps forgotten who the captain of the ship is? The management should *always* be held totally *accountable* and *liable.* The safety department is at times forced to juggle the injury figures because of this responsibility put on their shoulders. This is the cause of many safety cover-ups.

Quality

Legislation does not typically cover items such as poor quality, management ability, and skills. This means that only the tip of the safety iceberg is regulated by legislation. In fact, legislation has become so intent on focusing on injury-producing accidents that the only time an employer is inspected is when the number of injuries exceed a predetermined level. As already discussed, this creates a complete false picture, as the fudging of injury records, misreporting, and underreporting do occur.

No Fun

Conforming to safety legislation is no fun. It appears to be a drag, and there is always this threat of punishment hanging over the organization's head. Fines are horrendous. One of the largest fines in U.S. history was a mine that was fined in excess of $1 million as a result of an accident that caused three fatalities. Normally, a severe blow such as an astronomic fine does not even jolt management and the organization into safety action.

Challenge Is Fun

Legislation that promotes voluntary compliance to the safety system and standards is legislation that is moving in the right direction. Managers enjoy a challenge. If safety becomes a challenge, it becomes fun. A structured safety system incorporated into the day-to-day management system that lays down standards of conformance and degree of achievements as management goals makes safety exciting. Such a system will ensure that legal compliance is automatically achieved.

Voluntary Participation

In a number of countries, safety consultants, safety organizations, and safety training institutes work very close in conjunction with the legal compliance agencies. Legal compliance agencies, rather than fining companies for nonconformance, refer them to safety organizations and consultants. These entities then approach the organization and, on a voluntary participation basis, train their employees and institute various controls to create a safer, healthier workplace. This method has proved very successful. It is hoped that proposed legislation in the United States will lead to a similar setup. The rapport between the external, unbiased consultant (or organization) and the management of the company is normally very good. Open discussions can be held and joint action plans compiled to ensure legal compliance.

Such proposed legislation will benefit an organization in such a way that should a safety organization or a consultant audit the organization's safety system and give it a clean bill of health, agency inspectors will not inspect that workplace for the ensuing twelve months. What a motivator! What a wonderful system! Why has it taken so long for the blinders to be lifted? This has been the success of the 5-Star Health and Safety Management System for more than thirty years.

Summary

Although safety legislation compliance does improve safety, it does not and should not replace a comprehensive voluntary safety system. Such a system makes it easy to comply with safety legislation. By following the guidelines offered by a structured safety system, achievement of legal requirements is almost automatic.

20

Paradigm 19: Safety Costs Money

PERHAPS THE BIGGEST FALLACY in the safety scenario is that safety costs money. It is often said that safety gives no return, is nonproductive, and is really a burden one could do without.

A well-organized and efficient safety department can save organizations money. They may not be able to maximize profits, but they can minimize losses and contribute greatly to the bottom line of the organization.

Costs of Nonsafety

The National Safety Council (1998) reports, "The total cost of unintentional injuries in 1997 was $478.3 billion, which includes estimates of economic costs of fatal and non-fatal unintentional injuries together with employer costs, vehicle damage costs, and file losses" (2).

The report stated that accidents at work cost the U.S. economy $127.7 billion for the year 1997. Motor vehicle accidents alone cost $200.3 billion. The 2005–2006 edition lists the total economic impact of fatal and nonfatal unintentional injuries as $574.8 billion for 2004. The report states that there were 99.9 million visits to physicians' offices as a result of accidental injuries during 2003 (2). Although these statistics are on a national figure, what do you think your accidents are costing your organization?

Workers' compensation and insurance to cover damage and so on are items that create a false sense of economic security. More money being drawn out of workers' compensation means higher premiums in the forthcoming years. As with all insurance, the more you draw out, the more you pay in. Because these costs are often hidden and intangible, managers seldom get an accurate bottom-line figure.

Safety Department

The safety department, if structured and manned correctly, can save the organization a lot of money. The advice and guidance given by this department should be taken seriously. I recall an incident where the safety department had carried out a risk assessment of a particular process. The major risk was highlighted, and control steps were proposed to management in a written report. Needless to say, this report was ignored, and an accident occurred. Although nobody was injured, the property damage accident created a major business disruption and loss of equipment and ended up costing $700,000.

Intangible Savings

The savings brought about by safety do not come in the mail in the form of checks. The benefits and cost savings are normally intangible, but they are there. One cannot emphasize enough the benefits of reducing all forms of accidental loss. Medical expenses are horrendous, and if most organizations do a costing, they will be staggered. These are merely the direct costs of accidents. Accidents also have the indirect costs that are very difficult to pinpoint. They could include product defects, employee morale, union and management relationships, and the organization's relationship among the local communities. Companies also develop reputations as a result of their safety efforts and failures.

Totally Hidden Costs

The toughest aspect of accidental costs is the totally hidden costs. These costs include items such as the impact on the family of the injured person, the adverse publicity gained by the organization, and the general effect felt by the organization as a result of the accident. In one organization, a fatality occurred. This fatality impacted not only the workforce but also the local community. Both were literally in mourning for more than two weeks.

A loss not prominently evident is a certain amount of resentment that is always leveled at the organization after a fatality. Totally hidden costs can also include employees' loyalty, dedication, and enthusiasm toward the organization. Often these cannot be costed, but they become evident if safety is not taken seriously.

Fines

Failure to comply with legislation will eventually catch up with the perpetrator. Today, astronomic fines are being leveled at organizations that do not comply with workplace safety legislation. Examples follow:

- An auto parts manufacturer that had experienced more than sixty incidents of crushed fingers was fined $1.6 million.
- After the death of an employee, a fishing organization was fined $116,000 for failure to incorporate and use the lockout systems.
- An organization was recently fined for not maintaining the legal record of workplace injuries and illnesses. This fine exceeded $30,000.
- Job safety violations cost a contractor in excess of $200,000.
- Some thirty-four serious violations during one construction project cost two construction companies nearly $250,000 in proposed penalties.
- According to the Occupational Safety and Health Administration's (OSHA's) website, Phillips Chemical Co. in Pasadena, Texas, was facing $2.5 million in proposed penalties after an accident that killed a worker and injured sixty-nine.

The list goes on and on and represents only the tip of the iceberg. As indicated previously, once an inspector enters the premises for one violation, other violations and nonconformances seem to spring out of the woodwork.

In some cases, a fine is levied for a repeat of each instance of the violation. For example, if one machine is unguarded, it may carry a citation of $1,000. For every other machine that is unguarded on the same premises, an additional $1,000 is added to the bill.

It is good management to conform to the local safety and health legislation. Conforming to the requirements proves to be cheaper than nonconformance. Conformance to the basic safety and health legislation is the foundation of a good safety system but is not necessarily a comprehensive and all-encompassing safety management system.

Most Cited Standards

According to OSHA, fines for the manufacturing industry for the period October 2004 through September 2005 are as follows:

Standard	Number of Citations	Penalties
The control of hazardous energy (lockout/tag out)	3,563	$2,951,943
Hazard communication	3,332	$732,635
General requirements for all machines	2,660	$3,491,214
Respiratory protection	2,476	$901,148
Wiring methods, components, and equipment for general use	2,087	$1,008,083
Powered industrial trucks	1,882	$894,156
Mechanical power-transmission apparatus	1,852	$1,158,265
General requirements	1,517	$875,854
Woodworking machinery requirements	1,140	$594,668
Abrasive wheel machinery	1,123	$472,126

Top of the Hit Parade

According to OSHA, for the period October 2004 to September 2005, the standards that were most often cited are as follows:

1. Hazard communication—7,600 violations
2. Fall protection—6,400 violations
3. Respiratory equipment—4,500 violations
4. Lockout/tag out—4,200 violations
5. Powered industrial trucks—3,300 violations
6. Wiring methods, components and equipment for general use—3,200 violations
7. General requirements for machines—3,000 violations
8. General requirements—2,400 violations
9. Ladders—2,400 violations
10. Mechanical power-transmission apparatus—2,100 violations

The 2000 figures are as follows for comparison:

1. Hazard communication—7,000 violations
2. Scaffolding—6,700 violations
3. Fall protection—4,000 violations
4. Lockout/tag out—3,800 violations
5. Respiratory protection—3,700 violations
6. Electrical wiring—3,000 violations
7. Machine guarding—2,800 violations
8. Electrical system design—2,200 violations
9. Power-press safety—2,000 violations
10. Power-transmission safety—400 violations

These violations are for *all* industries. Highest violators in the manufacturing industry were lockout/tag out, followed by hazard communication and general machinery requirements. Little has changed since 2000, when the top of the charts for the manufacturing industry was hazard communication, followed by lockout/tag out and machine guarding.

Arising from more than 28,000 OSHA inspections, 114,500 citations resulted in fines amounting to more than $103 million for the period from October 2004 to September 2005. It is a staggering thought: $100 million worth of penalties for noncompliance to basic safety and health laws.

In the construction industry for the same period, general requirements hit the top of the charts with 9,500 violations. Fall protection was next with 6,400 violations, ladders with 2,300 violations, followed by excavations with nearly 2,000 violations.

Safety as a Value

Integrating safety management into the day-to-day management of the organization can save money. Saying that safety is first but not putting it first will not save the organization money. Management must put their money where their mouth is and make safety an organization value. This would entail creating a safety culture that permeates from senior management down to lower levels.

Safety should not have to be policed and monitored by legal and government agencies. Good safety is good business practice. Abiding by legal standards and requirements drafted to protect employees from accidental injuries and death should be used as the minimum safety standards for any workplace.

Safety should become an organization's *value* rather than a *priority.* "Safety is a priority" is a paradigm because priorities change. Values are ingrained and remain for a long period as part of the culture. Implementing a good structured safety system can help reduce the direct, indirect, and totally hidden costs of unintentional losses. The initial time and money invested will pay off in averting numerous and sometimes astronomical fines.

As one senior manager put it, "If you think safety is expensive, try an accident."

21

Paradigm 20: Discipline Makes People Work Safely

SAFETY SELDOM REALLY GETS THE ATTENTION that it deserves. The prevention and control of accidental losses is often put on the back burner until something happens. When something happens and there is an injury or potential for serious injury, then discipline is resorted to as the cure-all for safety violations. This is often an emotional reaction to situations that are often not the person's fault.

Discipline is why safety is still lagging in industry. Discipline has made safety the enemy of the common working man. As soon as there is an injury and an unsafe act can be proven, the person receives some sort of punishment.

A good friend and colleague of mine who represents a number of unions at a particular work site could summarize this chapter by his statement to me: "Discipline; it's never worked for 40 years! Why will it work now?" This summarizes how poorly discipline fits into safety.

I recall the case of an employee who was injured in an accident. Regardless of the causes of the accident, he was told to write an essay during his convalescence explaining what he had done wrong and what he would do in the future to prevent a recurrence of this accident. The irony of this matter is that his injury was the result of another person's unsafe act. Yet the injured person received the discipline.

Discipline is what forces true safety to go underground. Discipline forces people to hide injuries. Discipline creates a climate in which people refuse to report near misses, which are the vital clues to preventing accidents. Near misses are the foundation of accident prevention.

Three Days, No Pay

In some cases, if there is an injury as a result of an accident, the employee who happened to have committed an unsafe act is normally sent home for three days with no pay. This disciplinary action is intended to prevent a recurrence of the unsafe behavior. I've interviewed a number of these victims and can assure you that, according to them, this was the most degrading and humiliating occurrence in their adult lives—the very shame of having to go home and tell your wife and children (maybe even mother-in-law) that you were unsafe at work and were being sent home for three days and three nights, because of your "actions." I find this ridiculous and certainly not fitting of modern management techniques in the twenty-first century.

The people I have interviewed, without exception, had silently sworn that they would never report unsafe acts, unsafe conditions, near misses, or injuries if they could possibly avoid it. They would do anything to avoid this humiliation. "Utter, utter humiliation" was their description.

Were they safer workers when they got back to the plant? No. Every one of them told me that they had become foxier, cleverer, and slyer. When questioned, they explained that they would merely keep injuries and accidents hidden. In the future, if there were an accident investigation, they vowed that they would concoct stories for self-protection and the protection of their fellow workers. We still use these inaccurate accident investigations to compile injury statistics on which we base future safety action.

The safety management principle of *definition* states that, unless the true or basic cause of the problem is identified, solutions cannot be forthcoming. Merely disciplining people does not fix the problem. Some organizations immediately test an injured person for drugs after an accident. If positive, the employee is automatically fired, even if the presence of drugs had nothing to do with the cause of the accident. He is found guilty without trial.

Warning Letter

This is a true story told to me by one of my students during a training course. During an incident recall session, this young man stood up and emotionally told the story of his accident. He was on the night shift when sometime during the shift he crossed between two heavy vehicles, one of which reversed and inadvertently crushed him badly against the other vehicle. He was rushed to the local hospital, but the surgeons did not have much hope for his survival. His intestines had been badly damaged by the accident, and he was moved to intensive care, where they were certain he would eventually die.

This young man, however, fought off death. He told how, on the very brink of death, he opened his eyes to find his bed surrounded by twelve of his work colleagues. Among them was his general manager. In recalling this story, he said that on seeing his general manager, he got the feeling that his colleagues and management really cared for him and were sympathetic toward his injuries. The general manager then produced and presented him a letter of warning for having taken the shortcut that caused the accident leading to his injuries. The letter stated that should he be involved in any further accidents, he may well be disciplined.

The young man survived the ordeal and returned to work about three months later. He never fully recovered and was often taken from work to the hospital in extreme pain as a result of his complicated injuries.

Shortly after sharing this story with us, he was flown overseas to specialists, as he was still having trouble with his digestive system and internal organs. He admitted to us that he should not have taken the shortcut between the vehicles but nevertheless questioned the entire system of his being disciplined for his actions, especially while he was lying in the intensive care unit close to death.

Discipline for safety violations is often used as a cop-out and a cover-up for poor leadership. Disciplining people seems to say, "I really care about their safety." This is not true. Disciplining a person is really treating the symptom and not the cause.

Another reason why safety is a cop-out is that while unsafe acts are committed and nobody is injured, things are fine and well. Often unsafe practices are condoned. As soon as there is an injury as a result of an unsafe act, discipline kicks in. This is what is called a traditional *safety knee-jerk reaction* to a systems breakdown.

Ask this question: Can you recall when last a supervisor or manager was disciplined because of an injury that occurred? Your answer is likely to be "never" because discipline always seems to focus on the employee and not the leadership. Think for a minute. Could the line manager have been in the wrong or not exercised the correct leadership? If the answer is yes, was he disciplined? Nine times out of ten, the answer is no. He gets off scot-free.

If we talk discipline, why discipline *after* the accident? The acid test, as far as discipline is concerned, is the following: if management disciplines an employee for committing an unsafe act, are they absolutely sure that no unsafe condition, not one, exists within their area of control? If that manager's work area has one flaw in it, how can he then discipline an employee's behavior for having a flaw?

Few if any of the line management I have worked with knew the difference between positive and negative discipline, so they always resorted to the negative kind. One of our divisions experienced tremendous and obvious resistance to the safety effort from a longtime employee. Nothing could change this man's attitude or encourage him to participate in the new safety system. The general manager of the division sent him to a regional safety and health congress along with the safety and other staff. Something happened to him at that conference. He returned a true safety disciple. His attitude was completely changed, and he won the monthly Safety Man of the Month competition for the next three months in a row. This was an example of winning the man over by positive discipline. Getting him to mingle with other safety conference attendees and hear from the speakers that safety was important in other industries really helped to convince him.

Unsafe Act—So What?

"But George committed an unsafe act and he got injured!" So what? The fact that George committed an unsafe act comes as no surprise. People commit unsafe acts all the time. Let's look at ourselves for a minute. Did we, during the past week, exceed the speed limit at any time by one or two miles per hour? If the answer is yes, then we too have committed an unsafe act.

The punishment should be leveled not at the degree of consequence, which is fortuitous but rather at the act itself. What should be identified is the reason why the person committed the unsafe act. Why did George do what he did? Why? Why? Why?

The trouble with using discipline as a "fix-all" remedy to safety violations is that it is based largely on emotions. Supervisors, leaders, and management get emotional when people are injured. When there are multiple injuries, they are inclined to get more emotional, and their decisions are based on emotion rather than actual fact.

Case Study

To clarify this statement, let me quote an example. A supervisor approached me with a four-page accident investigation form and said he needed help with an accident investigation. What had apparently happened was that there had been a cave-in underground and three people had been buried up to their waists in material. They were rescued with minimal effort and suffered no injury whatsoever. The potential for serious consequences, however, was very high. This cave-in could have buried and suffocated them.

I took the accident investigation form from him and reviewed it. The front page, listing the names of the employees involved and the description of the accident, was half completed. The second page, listing the unsafe acts and unsafe conditions, was totally blank. The basic causes had not been identified, nor had any other part of the form been completed. The only completed section on the form was the remedial measures listed under the action plan on the last page.

Step 1 of the remedial measures stated that these three employees would be disciplined. I then asked the supervisor if he had completed the investigation and identified the basic causes of the accident. He replied that he hadn't, but he was still adamant that the three people were going to be disciplined.

Remembering that "prescription without diagnosis is malpractice," we then conducted a thorough loss causation analysis of this accident. We found no fewer than seven unsafe acts and unsafe conditions. Further delving revealed fourteen basic causes that had created the unsafe acts and unsafe conditions. These, in turn, resulted in the very serious, high-potential accident.

Even after sharing this information with the supervisor, he nevertheless went ahead and disciplined the three employees for their role in the accident.

A very simple test is to ask the question, "Will the remedies solve the problem?" If not, why do it? Remember to treat the causes and not the symptoms. In safety, we are constantly treating the symptoms and ignoring the real causes. Our safety paradigms make it easier to avoid the true issues. Blame the employee, find only the fault and not the facts, and safety will never improve.

Proof

If discipline has played any part in improving the safety of people at work, I challenge anyone to bring me proof of this. All discipline does is drive the accidents and injuries underground. Hundreds of employees have told me, "If I'm injured and I get punished, I will not report my next injury!"

We had the case of an employee who slipped, fell, and fractured his right arm at work. He asked his supervisor's permission to visit the doctor. Because he feared disciplinary repercussions, when questioned about his injury, he told the supervisor that he had been playing tennis the day before and had slipped and fallen on his arm and suspected that it was broken. He was treated at the hospital and returned to work, having paid for his medical treatment with his own insurance. No work injury was recorded, no disciplinary measures were levied, and, most important, the safety record was not spoiled.

The challenge here is open. Is there any proof that discipline has improved safety in the workplace? How is safety measured anyway? If only by the number of injuries, then I guarantee that safety has not improved one bit. In fact, it has probably worsened. What has happened is that discipline has driven the truth (the real number of injuries) underground, and the organization is suffering from the "ostrich syndrome." We think we are doing well, but people are still being injured. Discipline is no solution, whatsoever, to any safety problem.

When last was a supervisor disciplined for running a pigsty? When last was a supervisor disciplined for condoning unsafe practices? When last was a supervisor disciplined for not rectifying an unsafe condition, such as an unguarded machine, a cluttered walkway, or poor stacking and storage? If this has not happened, then surely safety is a cop-out used to beat people on the head.

"I Would Rather Fire Him . . ."

This is perhaps the greatest emotional cop-out that I have heard in the safety profession. Numerous supervisors and foremen have said to me, "I would rather fire him than have him killed on my shift." Or the emotional blackmail begins with, "I could never face up to having to tell his wife that he was killed because of his unsafe actions."

If a person commits so many unsafe acts or creates so many unsafe conditions, why did you employ him in the first place? Was the company not at fault initially? If you could not direct the habits and activities of such a person, was the leadership skill perhaps lacking? Firing a person confirms failure to manage that person.

As stated earlier, managers are in the toughest business in the world: *managing.* Safety management is the second-toughest business in the world. Firing people left, right, and center will not improve safety. Emotions should not cloud management's decisions. Emotions should not get people fired. There is no such person as an accident-prone person. The term *accident prone* is yet to be proved or disproved, and my advice is to remove it from the safety agenda entirely, as it is another safety paradigm.

Safety Policeman

Members of the safety profession who police safety belong in the security profession and not the safety management profession. Safety cannot be policed. Safety should not be forced on people. Safety should be a matter of guiding, educating, training, and motivating people to work safer. I often hear people say, "But disciplining is not always negative. There is positive discipline where you coach and council the person." Rubbish. That is warm and fuzzy and does not exist in the real world. Discipline in safety is almost always punitive.

Viewpoint

In one of the most controversial articles I have ever read, the author wrote an article about what happened to discipline in the workplace. He heads this article by making the statement, "It makes employees accountable for their actions." Just reading the introduction line almost made me want to laugh. Make *them* accountable for *their actions*? What a cop-out. What about *management* being accountable for *their* actions? The article continues in a negative vein by stating that "verbal reproof is an encouragement to be responsible." I find this ridiculous. Verbal reproof is what will drive all of your honesty in safety underground. The article continues, "Finally document your verbal reproof in case the employee pushes the limit." Push the limit? In other words, is the writer scared that the employee may start opening a can of worms by telling the truth? The truth may be that this is the way he has been encouraged to do whatever he's been doing, whether rightly or wrongly, for a number of years.

The author continues discussing almost school-yard punitive measures. He says that isolation, restrictions, and loss of privileges are also corrective actions. He rambles on, "Let's say you're conducting safety meetings and you have a particularly disruptive employee. What do you do? After the safety meeting, call the person aside and inform him that he is to fulfill his monthly safety training alone, without the group."

This sounds very much like the type of punishment that was dealt out at junior schools in the 1950s. This reminds me of the naughty child who has to sit in the corner with a dunce cap on his head. Isolation and loss of privileges—is this really where the safety profession is going? He continues this ridiculous argument by stating, among other things, that chastisement is used to correct consistent bad behavior or rebellious activities. Further insult is added to injury by his statements that the goal is to bring employees back into obedience.

He continues to give advice: "But if employees continually break safety rules, our job as safety professionals is to find a way to discipline them, or if required, fire them, before they hurt themselves or others—and then blame the company for whatever problems they caused."

The article concludes with the author stating, "I really don't care if employees like me. I wasn't hired to be liked, but to keep employees from getting hurt and to reduce the number of injuries and illnesses."

Well, I have news for you Mr. Safety Cop, if you are hired as a safety manager as the article indicates, then I suggest that your position title be changed. If you were hired to keep employees from getting hurt, then you took a job that has made you responsible for safety. Safety personnel should never be made responsible for safety. Responsibility for safety and the reduction of injuries and illnesses is management's job. You should be motivating, guiding, and training employees in the *techniques* of safety management rather than bullying them. You should not personally be taking on the responsibility of reducing the number of injuries and illnesses. Face the facts. All that is going to happen is that the employees will hide the injuries and illnesses from you. The safety reign of terror has never worked and will never work. That's another myth.

I will not quote any more from this article, which I find to be derogatory to the safety profession. My personal message to the author is that if he wants to fire people and discipline people, then they had better supply him with a security officer's uniform and a baton. Safety cannot be policed. Employees should not be policed. Firing people and disciplining them for safety is a poor excuse for proper management. The safety profession's objective, as I remember it, is to guide, educate, train, and motivate employees to prevent accidents. This cannot be achieved by disciplining and punishing them.

A colleague of mine had the following sign posted on his office notice board. The misguided message and concept of a safety person's function read as follows: "I'm here to save your tail, not to kiss it!"

Summary

The point of this chapter is that discipline has no place in safety management. I believe that a manager should lead people and not push them. I find it hard to understand that an employee can be punished for a safety violation when managers are guilty of numerous safety violations in the form of unsafe acts and unsafe conditions within their own work area, yet *they* are not punished.

If safety professionals have been appointed to reduce the number of injuries or the lost-time injury frequency rate, severity rate, or disabling injury incidence rate, they should resign now. If not, they are making a statement that in their personal capacity they can reduce injuries. This is impossible. Safety professionals have neither the authority nor the means to make this happen. By disciplining people, they become police officers and will achieve nothing. Line management is, always was, and always will be responsible and accountable primarily for the safety of the employees of the organization. After all, they have the authority. If that authority fails to guide people to follow the procedures and to tow the safety line, then other measures are called for, not before. If there is no structured procedure and steps to follow as a last resort, then once again the system is questionable. The myth that discipline improves safety is just that: a myth. Discipline is a cop-out for poor safety management.

22

Other Paradigms

IN SAFETY, THERE ARE A NUMBER of other paradigms that also hide and distort the real reasons behind accidents and that render a lot of effort by a lot of people useless. They are also ideal as safety cop-outs for safety responsibility and accountability.

It'll Never Happen to Me!

This is a very common belief that most of us at one stage or another have. We seem to think we are beyond being injured in accidents and carry on with certain bravado regardless. It'll happen to others, but I will never be involved in an accident. The truth is that none of us are safe. Life has certain risks that we are exposed to every day. Preservation of self and family depends on determining the risks of all life situations and deciding whether the taking of the risk is worth it. Driving is an example that affects us all and, with thousands dying as a result of road accidents every day, is a clear indication of the risk involved each time we climb into our vehicles. Since there is already risk involved, we should not push our luck and increase the risk by exceeding the speed limit or by taking some other unsafe action. We should rather drive with a defensive attitude and take defensive action. Is taking a chance on the road worth the benefit?

Saving Time Increases Risks

As a teenager, one of my heroes was a young man who had driven 400 miles to the coastal resort of Durban, South Africa, in less than seven hours, apparently twenty minutes faster than the previous holder, the driver of an MG sports car. This young guy had taken all the risks possible to beat the existing record, and he was hailed as some sort of a hero among the teenagers.

At a party one Saturday night where all the local youngsters hung out, the hero was constantly being congratulated and patted on the back by his pals, myself included, for this great achievement. Halfway through the celebrations, an old man walked up to the hero and also shook his hand. While doing so he said, "Well done, young man, I believe you beat the record by twenty minutes." He then said, in all sincerity and with the wisdom of his years, "So tell me, what did you do with the twenty minutes that you saved?" The young hero and the entire crowd were silent for a moment as these words sank in. We suddenly all got the message, as the hero was struck absolutely speechless by the question and had no answer. We then realized the impact of the question, and it dawned on me that to beat the record by twenty minutes, some hair-raising driving had to be done with the accompanying risk taking. The risks were to both the driver and other road users.

I learned one of the biggest lessons of my life that night: Is the end accomplishment worth the risk? Is it worth it? What if something goes wrong? Could you justify the consequences against the actions? My answer was no.

But We've Tried That Before

Many organizations have this common resistance to change. They have tried everything to prevent accidents in the past, and it didn't work. When questioned, they will tell you they have tried to motivate employees to work safe, they have attempted to get near misses reported, and they have done all in their power to make the workplace safer. Nothing seemed to work. That's a myth. What has probably happened is that a halfhearted attempt was made but they had given up because of the difficulty of the task.

To change the culture of an organization is a very difficult task indeed. It takes a lot of patience, tenacity, guts, and drive. Many people within the organization would like to see the old culture remain. This may lead to the tremendous resistance often experienced during a safety culture change process.

Resistance to Safety Change

Instituting safety culture change into an organization is similar to riding a bicycle up a hill. The minute you stop peddling, you go backward. To change culture, therefore, keep on peddling no matter how steep the hill. When introducing positive, proactive safety systems and actions, persevere to the fullest and do not let the actions be sidetracked.

One of the reasons many fail in their attempt to change the safety culture is because they try to change the culture from the bottom of the organization. This has never and will never work. Executive management must take the lead and set the pace for change. Without change leadership from management, the change will not happen. Mostly, the biggest resistance to safety change comes from executive management.

When I was brought in to effect a major safety culture change in an organization that had done everything to reduce the injury rates, I met up with resistance from all quarters. One of the good systems they had in place were safe operating procedures, called Job Safety Practices, or JSPs. Unfortunately, they had written one for almost every single task on the forty-acre plant site. The old culture called for the safety manager to sign off each one as being safety correct. This meant signing literally thousands of these JSPs every year. I read through some and found that I did not understand the procedures and therefore could not put my signature to the documents certifying that they were safety correct.

Meanwhile, the managers were sending me hundreds of these documents for signing so that they could meet their annual objective of having reviewed their JSPs. Under pressure, I took one procedure and read it thoroughly. A job safety analysis should always list the steps of the task in a logical sequence. Well, this JSP that I was scrutinizing was completely out of step. I remember clearly the step where the employee had to climb into the transformer that was being reconditioned. After he had climbed into the transformer, the next step read, "Drain all the oil from the vessel before entering." My following of the procedure showed that the man already in the oil-filled transformer was now being told to drain the oil. But it was a bit late for that now.

On review, I discovered that a number of the JSPs were totally incorrectly sequenced and sometimes put the worker at risk. I was relieved that I had refused to sign even one of them. I highlighted a few blatant mistakes on some of the procedures and sent them back to the originators for correction. At the same time, I attempted to get the wording of the JSP policy changed in order for the manager of the section and not the safety department to sign off on the procedures. A week later, I got a message from my immediate report, the deputy chief operating officer, that according to the executive I was "not a team player." It took a long time and sheer determination to eventually get management to accept responsibility for their JSPs.

The Principles of Resistance to Safety Change

The two most applicable safety management principles are explained as follows:

The safety culture that is embedded in the organization tends to prevail in the future. Safety attitudes and behaviors experienced in the past tend to be carried over to the future. Safety management normally experiences more resistance than any other aspect of an organization's business. The introduction of safety standards and procedures that differ from the way things were done in the past tends to be met with resistance by the people involved.

Introducing a safety management system is a change that creates an insecure environment. Introduction of safety programs requires adequate preparation, and the best way to introduce a comprehensive safety program is by introducing it element by element.

The principle of reaction to safety change is that the smaller the changes, the smaller the resistance to the change.

Management Need Not Take the Lead

Thinking that a safety culture can be created without the leadership and commitment of executive and all other levels of management within the organization is a myth.

Remember that people normally pay most interest to what the chief executive officer wants them to pay interest to. If this happens to be safety, the chief executive automatically sets the trend.

Does involvement of the executive come from being forced to comply to certain legislation, as a result of pressure from the workforce and the unions, or as a result of identifying and realizing the benefits of complying with accepted norms?

A chief executive and his fellow officers have two choices. They can either give minimum commitment to the safety movement or maximum commitment and support.

The initiative to launch and maintain safety momentum is required from top management, as is the leadership and ongoing support necessary to prevent downgrading accidents that lead to losses.

All managers should be visibly committed to the safety drive, and this visibility will involve their frequent visits to the shop floor to mingle with the workers and manage the safety at the point of action. The closer to the point of action they are, the more effective the safety management will be.

Interest in the safety of workers, full support to their safety requirements, and the ongoing supply of resources can be facilitated only from the management level.

Accidents Do Happen

Sometimes a difficult paradigm to explain is that accidents do not just happen but are caused by a series of small blunders coming together resulting in a loss. People sometimes believe that accidents, injuries, fires, and high-potential near misses seem to occur for no reason whatsoever. The truth is that accidents do not just happen. They are caused, and they can be prevented.

Those accidents that are as a result of the forces of nature are often difficult to predict and sometimes impossible to prevent. It could be argued that these accidents just happen, cannot be predicted, and cannot be prevented. This is true but is applicable only to a small percentage of total accidents. The majority of accidents are caused by a failure of some control system as a result of the risks of the situation not being assessed and minimized in time. The majority of accidents therefore are caused and do not just occur.

If we had to believe the myth that accidents simply happen, little or no effort would be made to prevent them from happening. One may as well give up.

Getting Hurt Is Part of the Game

To some macho mine workers as well as other rough-and-tough individuals at the workplace, getting hurt is all part of the game to them. Once again, peer pressure seems to dictate that getting hurt on the job is not such a bad thing and is par for the course. This mind-set is a paradigm, as we know that even the most hazardous jobs can be done successfully within the time limits without necessarily injuring anyone. An organization needs to get this message across to all who think that getting hurt is part of getting the job done.

Admittedly, some jobs are more dangerous than others, but even then the necessary precautions need to be taken and safe methods to carry out hazardous jobs found. Experience has shown that even in the hazardous construction industry, if the company makes up its mind to integrate safety into the project as a priority, it normally results in a safer project.

When Your Number's Up . . .

I do not intend to trespass on any religious beliefs with this paradigm, but I must argue that if this is the belief, why tempt fate? By operating in an unsafe environment or by taking shortcuts and committing unsafe acts, one tends only to push one's number closer to the front of the line. If a person believes that the inevitable will happen, then it is difficult for an organization to convince them otherwise. However, the systems, supervision, and structure of the work process should be such that it does not allow for the worker to be unnecessarily exposed to risk. Some beliefs in parts of Africa are that if a limb is amputated in an accident, then the Gods were angry with the worker, and the injury was a result. A very difficult case to argue is that accidents are caused by factors and aren't simply things that happen to people.

Anyone Can Become a Safety Director

This is one of my favorites. I attended a lecture in Phoenix, Arizona, given by Dr. Mike Manning, author of the book *So You're the Safety Director!* (Manning 1998). On opening the lecture, he asked for a show of hands from the group to indicate who had applied and was selected for the job of safety director or something similar to the job they now held. I was amazed to see only six hands appear from the group of more than a hundred seminar attendees. He then asked who had been pushed into the position without having been asked or given a choice or without requesting the position. Most of the hands were thrust into the air.

With a shock, I realized that the majority of the safety directors and coordinators in the room had actually been forced into the safety job that they had neither requested nor applied for. Some had admittedly chosen safety as their career and had embarked on appropriate courses of study and continuing education, but for the majority, safety had chosen them.

It dawned on me then that safety could be used as a dumping ground for employees who were not wanted in other divisions of the company. This is possibly because of the paradigm that anyone can become a safety professional. As with most professions or careers, there must be first a desire, and then an aptitude, followed by the necessary skills and, most important, the formal education. Gone are the days of the safety officer beating people into safety submission. Safety is a science and should be recognized as such. Safety is also a profession with a professional language, professional membership organizations, and formal educational programs. Putting nonselected, unsuitable people into safety roles for any other reason than because they are suited for the role is a slap in the face of the profession.

Putting injured employees into the safety department for recuperation (and mental adjustment?) after the accident is also wrong. Is this done as some form of punishment? I don't understand the reasoning for this practice.

The Discipline Myth

DuPont's training manual titled *Managing Safety: Systems That Work for Operations Managers* (1997) states that between 5 and 10 percent of employees "do not respond to these positive motivators but persist in ignoring safety rules" (97). But this is not researched information. They add insult to injury with the following unfounded suggestion: "You will probably have to apply discipline" (97). (This uses fear as its main motivator.)

Injury-Free Paradigm

Other organizations teach employees that workplaces can be "injury free." This is a tough one and, if really analyzed, cannot become a reality. An accident-free or consequent injury-free workplace is another paradigm that, if looked at carefully, is impossible.

Conclusion

There are a number of safety myths that need to be exposed. Safety philosophy statements should be carefully examined before being believed. Safety professionals should examine in detail the real cause of accidents in their plants and be unbiased in their investigation findings. They should also be prepared to stand up for what is right in safety. Few people want to change these safety paradigms and prefer to use safety as a reason for employee blame.

Many calamities have occurred recently as a result of a lack of safety standards and control, and they have shown that the causes of these events are deep rooted within the organization and are not simply superficial misbehaviors of individual workers.

Changing paradigms that strangle safety efforts is something that is going to take time, guts, and determination. People do not like to hear the truth when it comes to safety, accidents, fires, or especially a fatal injury. Is this perhaps why safety has become the sham of the century?

References

American Society of Safety Engineers. 1999. *Principles and Canons and Proposed Professional Code of Conduct.* Des Plaines, Ill.: American Society of Safety Engineers.

Arizona Daily Star. 2000. "Airline Knew of Problems before Crash," June 25, 4.

BHP Engineering. 1998. "Safety Pays." *BHP Review.*

Bird, F. E., Jr., and G. L. Germain. 1992. *Practical Loss Control Leadership.* 2nd ed. Loganville, Ga.: International Loss Control Institute.

Bird, F. E., Jr., and G. L. Germain. 1996. *Practical Loss Control Leadership.* 3rd ed. Loganville, Ga.: Det Norske Veritas.

Buys, A. A. 1994. "Managers Have You Lost Touch?" *Alberton Newsletter*, 4–7.

Carollo, Russell. 2000. "Military Aircraft Unsafe." *Dayton (Ohio) Daily News.*

Columbia Accident Investigation Board. 2003. Report. Vol. 1. NASA. August. Available at www.nasa.gov/columbia/home/CAIB-Vol1.html (accessed July 2006), 49, 121.

Consumer Product Safety Commission. 1998. "Product Recalls." (Article originally published in *Professional Safety*, 1998: 11–12.)

Compensation and Working Conditions. Fall issue (1997).

Daily Courier. 1999. "Labor Dept. Fines Local Company," March 14.

DuPont. 1997. *Managing Safety: Systems That Work for Operations Managers.* Wilmington, Del.: E. I. du Pont de Nemours and Company.

English, William. 2006. "The Moral Component of our Professional Challenge." *Professional Safety*, July, 55.

Friend, M. A. 1997. "Examine Your Safety Philosophy." *Professional Safety*, February, 34–36.

Fulwiler, Richard D. 1999. "Safety First? No, Not Really." Available at http://penton.com/oh/member/library/HTML_articles/1997/1097/safetyfirsmag10_97.html 4/16/99.

Gilmore, Charles. 1970. *Accident Prevention and Loss Control.* New York: American Management Association.

Hawes, Rachel. 1999. "Former Insider Turns Flame on ESSO." *The Australian,* May 4, 5.

Health and Safety Executive. 2006. Available at www.hse.gov.uk/statistics/causinj/index.htm, July.

Heinrich, H. W. 1959. *Industrial Accident Prevention.* 4th ed. New York: McGraw-Hill.

Hoyle, Bill. 1995. *Fixing the Workplace and Not the Worker.* Lakewood, Colo.: Oil Chemical and Atomic Workers International.

International Labor Office. 1983. *Encyclopedia of Occupational Health and Safety.* 3rd ed. Geneva: International Labor Office.

Johnson, D. 1998b. "What Does Management Think about Safety and Health?" *Industrial Safety and Health News* (15th White Paper survey), December, 22.

Leney, Fiona. 1994. "Three Miners Dead, but Production Will Not Be Affected!" *The Johannesburg Star,* 14.

Manning, M. V. 1998. *So You're the Safety Director.* 2nd ed. Rockville, Md.: Government Institutes.

Manuele, Fred A. 1997. "Principles for the Practice of Safety." *Professional Safety,* July, 31.

Manuele, Fred A. 2004. "Is a Major Accident About to Occur in Your Operations?" *Professional Safety,* May, 22–25.

McKinnon, Ron C. 2000. *The Cause, Effect, and Control of Accidental Loss, with Accident Investigation Kit.* Boca Raton, Fla.: CRC Press.

National Safety Council. 1993. *Accident Facts 1993.* Vol. 1993. Itasca, Ill.: National Safety Council.

National Safety Council. 1997. *Accident Facts 1998.* Vol. 1997. Itasca, Ill.: National Safety Council.

National Safety Council. 1998. *Accident Facts 1998.* Vol. 1998. Itasca, Ill.: National Safety Council.

National Safety Council. 2006. *Injury Facts 2005–2006 Edition.* Itasca, Ill.: National Safety Council.

Occupational Hazards. 1999a. "DuPont Fined for Recordkeeping." *Occupational Hazards,* April, 20. (Reprinted with permission. Material is copyrighted to Penton Media Inc.)

Occupational Hazards. 1999b. "National Safety Survey." *Occupational Hazards,* December, 30.

Occupational Safety and Health Administration. 2006a. www.osha.gov/pls/imis/citedstandard.sic?p_esize=&p_state=FEFederal&p_sic=all.

Occupational Safety and Health Administration. 2006b. www.osha.gov/pls/imis/citedstandard.sic?p_sic=B&p_esize=&p_state=FEFederal.

Petersen, D. 1978. *Techniques of Safety Management.* 2nd ed. New York: McGraw-Hill.

Petersen, D. 1998. "What Measurement Should We Use, and Why?" *Professional Safety,* October, 37–40.

Petersen, D. 2000. "Behavioral Approach to Safety Management." *Professional Safety,* March, 37–39.

Petersen, D. 2005. "Setting Goals Measuring Performance." *Professional Safety*, December, 43–48.

Purdey, Grant. 1996. "Risk Management." In *Practical Loss Control Leadership*, 3rd ed., edited by F. E. Bird Jr. and G. L. Germain, 419. Loganville, Ga.: Det Norske Veritas.

Scannell, Gerald F. 1999. "Ten Obstacles to Workplace Safety." *Safety and Health*, January, 28–31.

Tarrants, W. E. 1980. *The Measurement of Safety Performance.* New York: Garland STPM Press.

Tourvest Retail (Pty) Ltd. 2005. *Health and Safety Policy.* Johannesburg: Tourvest Retail (Pty) Ltd.

Tye, James. 1996. "Measurements of Failure." *Safety Management*, September, 30.

U.S. Department of Energy. 1999. *Occupational Hazards.* Washington, D.C.: U.S. Government Printing Office.

Index

About the Author

Ron C. McKinnon, CSP, is an internationally experienced and acknowledged safety professional, author, motivator, and presenter. He has been extensively involved in safety research concerning the cause, effect, and control of accidental loss, near miss reporting, accident investigation, and safety promotion.

McKinnon received his diploma in safety management from Technikon SA, South Africa, and a management development diploma from the University of South Africa in Pretoria. He received a master's degree in safety and health engineering from Columbia Southern University in Alabama.

From 1973 to 1994, McKinnon was affiliated with the National Occupational Safety Association (NOSA) in various capacities, including safety and health training and motivation. He is experienced in implementation of safety programs, auditing safety systems, and production of training films and videos. During his tenure with NOSA, he implemented safety systems and did training in seven different countries.

From 1995 to 1999, McKinnon was safety consultant and safety adviser to Magma Copper and BHP Copper North America, respectively. At BHP

Copper, he was a catalyst in the safety revolution in the copper industry that resulted in an 82 percent reduction in the injury rate and an 80 percent reduction in the severity rate.

In 2001, he spent two years in Zambia introducing the world's best safety practices to the copper mining industry. Thereafter, he accepted a two-year contract in Bahrain, where he successfully implemented a safety culture change at the country's second-largest employer.

McKinnon is the author of *Cause, Effect and Control of Accidental Loss* (2000). He also wrote *Safety and Health at Work: An Introduction*, currently being reviewed for publication.

McKinnon is a professional member of the American Society of Safety Engineers, a past president of the society's Tucson chapter, and an honorary member of the Institute of Safety Management. He is currently a safety consultant, safety change agent, motivator, and trainer.